Beyond Space
Beyond Matter

Science in Nāsadīya and other Sūktams

Krishna Ramadas

ISBN-13: 978-1544223988, ISBN-10: 1544223986

BISAC: Science / Cosmology

I dedicate this book to my Guru Sri Sri Ravishankar who works tirelessly to sustain Vedic traditions while infusing scientific temper into the interpretation of Vedic mantras.

Table of Contents

Beyond Space Beyond Matter

Preface

The work on this book started when Sri Sri Ravishankar suggested that I present the contents of the Nāsadīya Sūktam and the Ambhasya-pāre Sūktam, considering the advancements of modern cosmology. Prior to this, I had researched in detail on the first translations of the Vedas by Europeans focusing on the historical narrative behind those translations. Sri Sri told me that European translators had worked with limited knowledge of the Vedas. He illustrated that by showing how many words in the Vedas convey deeper scientific ideas. The word parṇa, for example, means leaf in Sanskrit dictionaries. Sri Sri explained how the word also implies photosynthesis which takes place in a leaf. He encouraged me to explore along those lines.

The approach of looking at words per Sri Sri's illustration was new to me. Expertise on such methods had declined and had likely vanished a few generations ago. There are no publications on those methods. We only know today of a simple method which explains how Sanskrit letters were used until the fifth century to represent numbers. A few Samkhya and Tantra texts assign a meaning for each alphabet of the Sanskrit language. They are, however, lacking in detail. I could only use them to see how well SriSri's illustration would fit per their methods.

SriSri's talk on the Nāsadīya Sūktam in a gathering sometime back became the road map for my research. He had referred, on several earlier occasions, to one phrase in the Sūktam and the phrase's connection to the idea of Dark matter in modern astronomy. I had to find similar connections for other phrases in the Nāsadīya Sūktam. Each mantra of the Nāsadīya Sūktam, per my research, conveys an astronomer's model of the creation of our universe. Mantras, however, go one step further.

Two of the mantras in Nāsadīya Sūktam introduce the idea of consciousness entering creation. I would not have been able to expand on that theme without some expert guidance. SriSri had asked me to tie the ideas about space in mantras to Yoga Vaśiṣṭā. Yoga Vaśiṣṭā is an ancient text which speaks eloquently on the topic of consciousness and the mind. State University of New York has published a translation of this text through its press as a reference material for the student of modern psychology. This text was an invaluable asset to me in completing my research on the nāsadīya and the Ambhasya pāre Sūktams.

I did not initially see the connection between the Nāsadīya Sūktam and the Ambhasya-pare Sūktam. I found a continuation of ideas from the former into the latter upon researching a bit. The latter references two other important Sūktams, namely, the Hiraṇya Garbha Sūktam and the Uttara Nārāyaṇa Sūktam which bring completeness to the idea of creation per Nāsadīya Sūktam.

There are three aspects to preserving the Vedas per SriSri. Preserving the chants by learning, practicing and teaching them is the first aspect. Researching into the meaning of mantras with a scientific temper is the second aspect. Using the mantras and rituals for the benefit of mankind is the third aspect. Mantras were designed to cater to the third aspect. The Ambhasya Pāre Sūktam gives us an understanding of the third aspect of mantras. The Ambhasya Pāre Sūktam is also an ideal point to begin additional research on the topic of Devas and other beings such as the Gandharvas.

This book includes a chapter for each of the four Sūktams, namely, Nāsadīya, Ambhasya-Pāre, Hiraṇyagarbha and Uttara Nārāyaṇa. The book analyzes ideas related to consciousness in the Nāsadīya Sūktam in context of Yoga Vaśiṣṭā in a separate chapter. It includes two chapters to illustrate different ways to derive the meanings for words in Mantras.

My attempt to interpret the nāsadīya Sūktam against modern cosmology is not a total deviation from traditional interpretations. I started my research on the Nāsadīya Sūktam with the commentaries of Sāyaṇācarya, from the 14th century, using it to understand the traditional viewpoint on creation. I decided then to deviate a bit on the interpretation of a few phrases within the context of modern science. A commentary of Sāyaṇācarya is included as a reference. I hope those seeking to understand the Vedic view of creation find this book to be a good starting point. Researchers may appreciate the book for the scientific temper with which the content of the Vedas can be understood.

Beyond Space Beyond Matter

1. Nāsadīya Sūktam

The nāsadīya sūktam of the Rig Veda is named after the first two words in it, namely, 'nāsad āsīt'. These two words convey the idea that the material universe was not there at the time of dissolution. This Sūktam is therefore believed to describe the way in which material existence arose from a state of dissolution. The seven Mantras of the Sūktam are fascinating for their parallels with the Big Bang theory of the creation of the universe. In them are references to several ideas in modern cosmology such as the inflation of space, Vacuum Energy, plasma fluid filling a nascent universe, and the important role of dark matter in shaping the structures of the universe. The Sūktam is a concise introduction to the Vedic model of the creation of the universe. It introduces the ideas about Brahman and Maya which the Samkhya texts elaborate.

The Sūktam starts with "Beyond space, in the beginning, there was neither existence nor non-existence; neither birth nor death." It then introduces the emergence of a universe of pure energy. One phrase in the third Mantra, namely, "darkness that wrapped around darkness' makes ostensibly little sense without the further depth of understanding provided by modern cosmology. The resemblance of the fifth Mantra to the modern cosmological idea of radiation separating from matter is eerie. The Sūktam then ends with questions about the stability of the universe. These questions remind us of puzzles which modern science is grappling with.

Rishi Prajāpati paramēṣṭhi is the seer of this Sūktam. He introduces us to Vedic ideas about consciousness. Rishis and Yogis excelled in their understanding about the mind, its modalities, and states of consciousness. They could discover how the universe works by tapping into the hidden powers of the human mind. Rishi Prajāpati paramēṣṭhi makes bold statements about the emergence of cosmic consciousness and the consciousness of Devas. His statements provide a platform for bringing two seemingly divergent time lines together, namely, the manifestation of consciousness and the emergence of material universe.

The human mind cannot comprehend the state of final dissolution. Rishi Prajāpati paramēṣṭhi therefore guides us to imagine a state of dissolution. He begins the first mantra by negating what we commonly accept as the salient marks of existence.

ṇasadasinnō sadāsīttadānīm nasīdrajō nō vyōmā parō yat
Kimāvarivaḥ kuha kasya śarmannambhaḥ kiṃāsīd gahanaṃ gabhīram

A pair of opposite words *sat* and *asat* appear frequently in Vedic literature. They denote ideas of changelessness and change, respectively. A change (asat) is inferred with respect to a reference point. One assumes that the reference point (sat) is static and not undergoing a change. The word *asat* applies to all entities which undergo a change. Nature is full of changes. Everything in nature has the characteristic of *asat*. An observer is of the nature of *sat*. We can extend this to a cosmic scale. The cosmos is full of changes. Stars form and explode. Galaxies form and merge with one another. Earth and space are filled with changes (asat).

Rishi begins the Sūktam by saying changes were absent at dissolution. Material existence (asat) which constantly undergoes changes was absent. Beings (sat) who can observe changing phenomena were also absent.

There is an agent behind any change. Change agents are of the nature of Rajas. The cosmic change agent (rajas) was absent at the time of dissolution. The wind stirs up stagnant air. Energy in wind is a change agent. Atmosphere is in constant movement. It is filled with *Rajas* or energy. The word *rajas* is often used to refer to the atmosphere.

Energy is behind any movement or transformation. Energy was also absent in the state of dissolution. We can conclude then that Vacuum, or space devoid of energy, must have existed at the time of dissolution.

Space time of a physicist begins at the Big Bang event. Quantum physicists argue that matter defines space. Space does not exist if all matter is removed from it. Scientists have debated the definition of space for the past hundred years. Albert Einstein accounted for the contribution by the empty space in his equation for the general theory of relativity. Space and vacuum states continue to remain active research topics to this day. The Rishi of this Sūktam, however, says that space was absent at the time of dissolution.

Several Sanskrit words refer to Sky or Space. Each refers to a unique aspect of space. Each word is derived from a different root. The diversity of these roots suggests that Vedic thinkers contemplated on the idea of space and vacuum. We will later consider the derivation of several words referring to space in chapter 4. Nāsadīya sūktam uses the word *vyoma* in the first Mantra to indicate space. Let us understand the Rishi's choice of this word.

We look through the atmosphere at the sky. Atmosphere is thinly spread out matter and mostly empty space. Deep space also contains matter. Matter there is dispersed to an extreme level. Earth's atmosphere (vyoma) is a container. Space is a container also; it holds the entire creation within it. The word *vyoma* refers to the placid sky which has no limit and to deep space.

The Rishi says that there was no sky or space (vyoma) during dissolution. The Rishi uses the affix *para* (beyond) with the word *vyoma* to indicate something beyond sky and space; he means a different kind of space. We will look at this in a bit. The combination of the two words *vyoma* and *para* reappears in the last Mantra.

Entropy and matter are indestructible as per the Laws of conventional physics. Matter transforms into energy when an atom explodes; it never disappears. Matter disappears without a trace only in a magician's act. It is difficult for the ordinary human mind to accept the idea of universe cycling through existence and dissolution. Scientists who support cyclical cosmology are also a rarity.

Rishis envisioned cycles of existence and dissolution millennia ago. The Rishi of the Nāsadīya Sūktam acknowledges the common belief that matter cannot disappear. The universe only disappears from one's view per such a belief just as sugar seems to disappear when placed in water. The Rishi counters the possibility of a wrapper[1] hiding the universe

[1] Wrapper is a loose equivalent for the Sanskrit word āvaraṇa. A wrapper hides a gift from view. The idea here is that the universe was not in a state of hiding.

with a couple of questions. Where (kuha) does the wrapper (āvarīva) stay if all domains of existence have dissolved? What is the nature (kim) of the wrapper if it can be neither *sat* nor *asat*? What power (kasya śarman) protects the wrapper from dissolution?

The Rishi introduces another poignant word, *ambha*. It is impenetrable (gahana) and mysterious (gabhīra). It is prakṛti, or primordial nature which is the undifferentiated seed of the universe. The ocean illustrates the principle of ambha. It does not overflow despite the overwhelming inflows from the rivers of the world. It is noted for its depth (gahana), and it is inexhaustible (gabhīra).

Biological evolution began in the ocean. The cosmic version of the ocean is *ambha*. Both the sentient and the inert arise from it. Physical space manifests out of this primordial cosmic principle. This primordial source of creation was absent at dissolution. We will know more about the word ambha in the fourth chapter. Let us make note of the fact that certain words such as *ambha* can convey a lot more information than their dictionary meaning.

ṇa mṛtyurāsīt amṛtam tarhi na rātriyā anha asīt prakētaṃ ānīt avātaṃ svadayā tadēkaṃ tasmāt ha anyanna para kiñcanāsa

An earlier creation had ended at the time of dissolution. A certain terminating condition or a destructive power was needed to annul that creation. The terminating phenomenon that acts on everything in existence is *mṛtyu*. Death leads a being towards its end. A being's gross body then disintegrates and disappears. Death (mṛtyu) also had disappeared after taking the previous universe to a state of dissolution.

Death and deathlessness (amṛta) are mutually exclusive. A created object or a being thrives until touched by the fangs of death and destruction. Death is a riddle. Death appears remote to anyone even after witnessing several others die. A human being's faith in life is unshakable until the very end. Deathlessness appears to reign in the psyche of living entities. Deathlessness (amṛta) was also absent at dissolution.

Conventional measures of time are relative. Ancient conventions for time measures were based on cadences such as days and nights. Vedic civilization had developed an advanced method of time keeping. They tracked small units of time unit to units as large as a Kalpa (4.32,000 billion years). Time and its measures had no meaning in the state of dissolution.

Scientists once thought time to be independent of a reference frame. They realized by early 20th century that time cannot be separated from the dimensions of space, in a relativistic context. The timeframe in which an object moves in space depends on its velocity relative to an observer and on the strength of gravitational fields in the vicinity. Gravity slows the passage of time for an object as seen by an observer at a distance. Physicists treat Space and Time as a combined variable while creating universally applicable models for systems in nature. Spacetime, a combined variable, allows physicists to simplify the task of making the same model applicable to super galactic and subatomic observations. Physicists call this variable as Spacetime.

The General theory of relativity assumes that Time did not exist at the creation of the universe. Cosmologists who favor cycles of expanding and contracting universes also assign a discontinuity to Time at the Big

Bang. The idea of Time breaks in both cyclical and linear theories about the universe. The Rishi of this Sūktam summarizes this by saying time was absent during dissolution.

Nāsadīya Sūktam negates, systematically, every major concept that attests to existence. Rishi then addresses the topic of consciousness. Nāsadīya sūktaṃ addresses cosmic consciousness as the One. The One continued to exist. It inhaled (ānīt). It breathed without drawing air in (avātaṃ). It relied on its inherent power (svadhā) to breathe.

The concept of all pervading consciousness is a taboo for scientists. They acknowledge the presence of consciousness only in a sentient being. They associate consciousness to the life force in the body of a being. Rishis and Yogis introspected deeply on the connection between Life force and consciousness. They recorded their insights in Vedic and Yoga literature. Yoga Vaśiṣṭā is an ancient text which expounds on consciousness. It introduces thought provoking ideas in the form of a dialogue between a student and a master. We will see excerpts from the text in chapter 3. These excerpts expand themes which the Nāsadīya Sūktam only mentions.

Rishi does not attach a name to the One that existed. We can only assume that the Rishi is referring in this Mantra to consciousness. We will see proof for this assumption from other related Sūktams. Chapters six and seven present two more Sūktams which are closely related to Nāsadīya Sūktam.

Consciousness is independent of Life force. Life force arises from consciousness. The Rishi conveys this paradox by the sentence "It inhaled without drawing any air in because of its inherent power". A theoretical physicist can see a parallel in this to the different energy states of vacuum. Bizarre nucleation bubbles can arise in space because of such energy level differences in vacuum. Our universe began per Physicists as one such bubble. The Rishi expresses the same idea as the One who inhales without drawing air in.

Most scientists accept Big Bang theory which suggests a big bang beginning for the universe. Observational evidences favor this theory in a big way. Scientists agree that the tiny space within a bubble inflated within a fraction of a second (10^{-32} seconds) and became the vast space which is the observable universe of today. Inflation smoothened out disorders which existed before it. The universe became homogeneous after the inflation. Scientists attribute inflation to an unknown energy field. They call it the inflaton. Rishi calls the inherent power of the One as *svadhā*.

The inflation field bore fundamental particles of physics. It decayed in the process. A tiny bubble inflated within a fraction of a second to become the universe! The Rishi expresses the same idea with the phrase "The One inhaled (ānīt) without drawing air in (avātaṃ)!"

Plasma fluid which contains fundamental particles of physics filled the universe at the end of cosmic inflation. The universe, per Vedas emerged from a cosmic egg. It is difficult to explain the reason behind the existence of similarities between the two ideas. The Rishi says in the next Mantra that boundaries were not noticeable within the primordial fluid (salila). Scientists express the same idea differently. The universe was homogeneous at the end of the inflation period. Scientists attribute the symmetry in the large-scale structures of the present universe to that homogeneity. The following mantra presents a similar idea which relates to the largse scale structure of the universe

tama āsīt ṭamasā gu̱lhamagrē aprakētaṃ salilamā sarvagm idaṃ tuccyēna abhvapihitaṃ yadāsīt tapasas tan mahinā jāyataikaṃ

The Rishi conveys what happened next (agre). This (idam) creation was like a featureless fluid (salila). This fluid had no distinguishing marks (apraketa). It was indiscriminate. One can relate *salila* to plasma state of matter. Plasma state of matter is featureless. Electrical charges dominate plasma, the fourth state of matter. *Salila* is an ocean of energy.

Our universe contains clusters and superclusters of galaxies. They are in threadlike arrangements. Scientists are studying the formation of galactic filaments in the early universe. The effect of quantum fluctuations in the nucleation bubble magnified when it inflated. Initial anisotropies grew larger over time. The large- scale structure of the universe was determined long before galaxies began appearing. The filament like arrangement of galaxies was predetermined.

Eddies and currents form in water constantly. Their Boundaries are impermanent. Fluidity of water implies the absence of structural boundaries in it. The Rishi refers to a different fluid state in the primordial universe. He qualifies the state of fluid by adding the word *apraketam*. This word means that structures exist, but they are concealed. We will analyze the construction of this word in Chapter 4.

The Rishi suggests here that the blue map of this (idam) universe of today, was in that fluid (salila). We can understand this idea through two analogies. A seed contains the structure of a tree. A block of granite carries a sculpture within it.

Tamas was an important element at that stage. "*Tamas* enveloped *Tamas*". We can understand the meaning of the last sentence with the help of an important idea in modern cosmology. The word *Tamas* has several interpretations. Primordial nature (prakṛti) has three qualities (guna) one of which is *tamas*. It is of the nature of inertia. It is therefore the direct cause of inert material existence.

Matter is inert. Any inanimate object is of the nature of *tamas*. Let us apply an extension of this logic to an idea in modern cosmology. Matter is visible though it is inert. It participates in the evolution of the material existence. Dark matter however is invisible and does not participate in the evolution of atoms. The bulky and inert dark matter is an even better representation of unresponsiveness (tamas) when compared to ordinary matter. Let us further explore the connection between dark matter and the quality of *tamas*.

The visible structures and related phenomenon compose only a tiny portion of the universe. A major portion of the universe is invisible and filled with Dark matter. The most advanced telescopes do not detect this Dark Matter. Scientists presume of its existence because it exerts gravitational force on the structures which are visible in the universe. Dark matter neither absorbs nor emits radiation. Scientists therefore do not know much about dark matter. Dark matter, per physicists, formed alongside ordinary matter during the Big Bang. This dark matter influenced the grand scale web like structure of galaxies in the Universe.

Nāsadīya sūktam emphasizes the presence of darkness (tamas) early in creation. Something dark (tamas) coexisted with the primordial fluid (salila). The above Mantra begins with an affirmation of the presence of tamas. It mentions primordial fluid (salila) later. The future universe was in a fluid form enveloped in something of the nature of *tamas*.

We notice three distinct ideas in the Mantra. 1. Darkness (tamas) enveloped the inert (tamas). 2. All that we see now (idam) lay hidden (apraketa) within a fluid (salila). 3. Void (ābhu) surrounded the inert something (tamas) and the fluid in which the unborn universe rested.

The universe emerged from out of the three through enormous effort (tapas). Is any effort necessary to overcome vacuum? Vacuum is after all strengthless. Rishis wonder similarly. He adds the adjective (tucca) in front of the word void (ābhu). The word *tapas* can also be taken here to mean heat. Let us understand the theme in this paragraph by studying a corresponding idea in modern cosmology.

Primordial plasma lay clumped in over dense regions throughout the early universe per scientists. Plasma was not uniformly distributed. Black matter lay towards the center of these regions. Two forces opposed each other in each overly dense region. Gravity pulled the region towards the center. Heat from interactions between photons and Baryons in plasma pushed the region outward. Every region began oscillating throughout the universe because of the opposing forces. Scientists call this oscillation as the cosmic sound. They assess the impact of the cosmic sound on creation by studying the structures of the present universe. Cosmic sound had resonated in the universe for a major part of the universe's plasma phase which lasted for 379,000 years.

The fluid universe struggled to break free. Nāsadīya sūktam refers to this struggle by the word tapas. A mountain has the power to weather a mighty storm. The universe is mightier than the biggest mountain by several orders. The Rishi wonders how something so powerful was constrained by something as insignificant (tucca) as void (ābhu). Whatever emerged after a long period of struggle (tapas) manifested in unison. We find a similar theme in modern cosmology.

Photons eventually escaped out from over dense regions of plasma. Electrons were free then to combine with nuclei to create atoms for the first time. Matter emerged in conventionally recognizable form concurrently across the whole universe. Matter popped up throughout the universe. Nāsadīya sūktam conveys this idea with the phrase "the One manifested (jāyata ēkam)".

We saw the phrase "The One inhaled without air (ānīt avātaṃ)" in the previous mantra. The One existed at the time of dissolution. Here we find the phrase "the One manifested". Is there a relation between these two entities? We will find a clear answer in later chapters when we understand the connection between nāsadīya sūktam and a few other sūktaṃs. Before that we need to understand the idea of mind.

Nāsadīya sūktam introduces the concept of mind (manas) in the next mantra. Consciousness differentiates into mind principle per next sūktam. This is a complex idea. Chapter 3 clarifies the origin of the mind by quoting from the Yoga Vaśiṣṭā text.

kāmastadagrē āmavartatādhi manasō rētaḥ pratamaṃ yadāsīt

satōbandhumasati niravindan hṛidi pratīśyā kavayō maniṣā

Nāsadīya sūktam narrates what happened next (agre). Desire (kāma) arose within the cosmic egg. It became the germ cell of mind (manas). Consciousness alone existed at the time of dissolution. There was, nothing for it to be aware of then. Thoughts arise when something comes in the purview of the mind. Consciousness existed during dissolution. Movements of energy had also come to a standstill.

Dissolution for cosmic consciousness is like deep sleep state for human consciousness. Thoughts vanish in deep sleep. The mind is only a field of thoughts and therefore does not exist during deep sleep. Energy movements arise in the field of prāṇa periodically during sleep. These create dreams. Dreams are thoughts. Mind becomes alive in dream state. Yoga Vaśiṣṭā compares existence to dream experiences of a dreamer. This viewpoint brings credibility to the idea of a close connection between consciousness and material world. We will see poignant elaborations of this idea from Yoga Vaśiṣṭā in Chapter 3.

Plasma state is purely an energy state of existence. Creation began when energy stirred up in vacuum according to scientists. The view point of Scientists and Rishis are in alignment regarding the primordial fluid. They both agree that it is energy centric in nature.

Cosmic egg per Rishis was filled with Primordial fluid (salila). Dynamic movements are a characteristic of this primordial fluid. The stirring up of energy in void created a reason for cosmic consciousness to focus on something else. It signaled the end of a state of total self-awareness. Nāsadīya sūktam calls the emergence of a new state as the manifestation of the One. Conditions were right then for thoughts to arise.

The very first thought to arise was a desire, per Nāsadīya sūktam. It was the germinal seed (retas) for the mind (manas). Cosmic mind could manifest the moment awareness moved outwards. A similar phenomenon recurs every morning when a sleeping person becomes aware of the surroundings. It is easy to see how some concepts about human mind and thoughts apply to the idea of a cosmic mind. Yoga Vaśiṣṭā expounds on this analogy. Chapter 3 contains several excerpts on the topic from Yoga Vaśiṣṭā text.

Human awareness which moves outwards every morning leads human consciousness into the world around. Cosmic mind similarly links the sentient and the inanimate. Nāsadīya sūktam therefore calls the universal mind as the kinsman (bandhu) of sentient (sat) beings. Sentient beings (sat) thus live within this inanimate (asat) material creation. The cosmic mind is the creator. We will explore this idea a bit more in later chapters.

The kinship of beings (sat) to the material universe (asat) is through only one link, the cosmic mind. This idea is not easy to comprehend. One's mind must drop its comfort with normal beliefs about existence. Skilled mind, per Rishi, can then recognize this connection. The question related to its own origin needs to arise in one's mind first. One's mind expands and becomes steady when understanding its own nature. One who voluntarily stills the mind is a seer (kavi). The intelligent ones (manīṣa) understand (niravindan) the truth about the consciousness connection. Such understanding arises from one's heart (hṛdi) after searching (pratīṣya) for it with a still mind.

Psychological conditioning keeps a mind outward focused. Such a mind is unable to find the connection to its source because of its tendency to look elsewhere. The truth is within one's heart. Chapter 3 explains the importance of dropping psychological conditioning with quotes from Yoga Vaśiṣṭā. The following chapters also present the role of one's heart as the seat of pure consciousness.

The human body including the brain is made up of inert chemicals. The brain, per conventional thinking, is however the source of sentiency in a human body. Sentiency appears within the brain when the body of a child is developing out of insentient molecules. Sentiency per usual understanding is short lived and disappears at death. Psychological conditioning of a human mind per Yoga Vaśiṣṭā begins with the above ideas. The root of awareness, per Rishi however is ancient. It is permanent because Consciousness supervised creation.

A theoretical physicist's understanding is close to the view point of a Rishi. Both agree that the creation began from a single source. The cosmic egg arose from consciousness per Rishi. Inflaton energy arose from a vacuum per theoretical physicist.

Consciousness filling Vacuum is the only point of divergence between the two. The Rishi does not disassociate creation from its source. Consciousness pervades creation. Inflaton energy however decays after creation per theoretical physicists. It plays a crucial role only in the early universe.

The Nāsadīya sūktam here returns to a thread from the third Mantra. The next mantra could as well be a summary of the Big Bang theory. The resemblance is unmistakable; it describes the universe moving out of its fluid form.

tirascīnō vitatō raśmirēśām adasvidāsīt uparisvidāsīt

rētōdā āsan mahimāna āsan svadhā avastāt prayati purastāt

Radiation (rashmi) tore away from the rest. It separated from the rest in a specific way. Radiation then cut crosswise (tirascīna) and spread out rapidly (vitata) like the predawn brightness. The space within the cosmic egg filled up quickly with brightness. We analyze the construction of the word *tirascīna* in Chapter 4. The word is important. It conveys the idea of emergence of radiation in the universe.

Photons escaped during the Big Bang, per scientists, and decoupled from charged atomic particles which were present then in the plasma fluid. This separation happened uniformly all over the universe. The universe had already expanded to its current observable limits by then. These photons became the microwave radiation which is detectable even today. Scientists have created a map of Cosmic Microwave Background (CMB) radiation. These maps help cosmologists model the evolution of the early universe. The universe took shape gradually from out of the initial matter which separated from photons.

The narration in Nāsadīya sūktam differs only slightly. Rudimentary matter (mahimāna) was not the only one which separated from the radiation filled. There was one other remnant, namely the germinal seeds of beings (rētōdhā).

The mighty forces (mahimāna) are building blocks of material creation. Samkhya texts list the number of mighty forces to be five. These are the five root elements of inanimate creation. The earth, water and air elements represent matter in three states. They correspond to solidity, fluidity, and movement. The fire element represents energy. The space element is the container principle. Objective space per Nāsadīya sūktam appeared only after radiation separated from the rest within the cosmic egg. Space and matter originated together. Scientists have arrived at a slightly different conclusion. Space is a feature of gravitational field. Space therefore cannot exist apart from matter and energy.

Brightness split away in a horizontal or crosswise fashion within the cosmic egg. What remained in the lower (adha) and the upper (upari) halves of the cosmic egg after the split? Nāsadīya sūktam raises this logical question to illustrate the split to be unusual. Germinal seeds of beings (rētōdhā) and mighty forces (mahimāna) remained back. They however do not correspond to an inferior and a superior half. They play equal roles in existence. One loses its significance without the other in creation.

Rishi asks a question but evades answering it. There must be a reason behind it. The word tirascīna provides a hint. Cosmic egg is a metaphor. Radiation did not escape from an egg which split open into an upper and a lower part. Radiation escaped all over within the cosmic egg and filled (vitata) the universe. Scientists explain the same idea differently. Photons were trapped within over dense regions of universal plasma. Charged particles such as electrons curtailed the movement of photons. Electrons however could pair up with nuclei at the end of the plasma phase. They no longer restricted the freedom of photons then. Photons emerged everywhere and crisscrossed each other. The Rishi conveys the idea of an escape from bonded existence with the word crosswise (tirascīna).

Germinal seeds of beings (rētōdhā) and mighty forces (mahimāna) were in a status quo before manifesting. The mantra identifies the status quo state as *svadhā avastā*. We saw the word *svadhā* in the second Mantra. It refers to the inherent power of consciousness. Samkhya text names the inherent power of consciousness (Brahman) as *māyā* which is of the nature of illusion. Yoga Vaśiṣṭā explains the nature of illusion in creation beautifully. We will see a few quotes from Yoga Vaśiṣṭā on the illusory nature of creation in Chapter 3. Germinal seeds of beings (rētōdhā) and mighty forces (mahimāna) were pushed up and forward (prayati parastāt) when radiation escaped. They were pushed up to a state of exertion (prayati) from an earlier state of comfort (svadhā).

Samkhya texts talk about the creation of mighty forces (mahimāna). Modern science describes the formation of matter as the photon epoch. There are many parallels between the two. Atoms are the smallest unit of matter in a conventional sense. Electrons are bound to a nucleus in an atom. Atoms appeared only at the beginning of the photon epoch. Electrons and nuclei moved independently all over the universe for a major part of an earlier epoch. This epoch lasted for 379,000 years and ended with the emergence of photons. Matter truly manifested when radiation emerged in the universe. Vedic texts consider matter to be one of the *mahimāna*. Mahimāna appeared when radiation escaped from primordial fluid.

Material creation started with the emergence of the mighty forces (mahimāna). The five elemental principles, namely, space, air, fire, water, and earth are also called as *mahabhūta* or *bhūta*. They mixed with each other in different proportion to create all materials and physical phenomena in the universe. *Mahimāna* are not in their elemental form in the universe today. Creation hastened when they began intermixing with each other. The next step of creation is *visṛṣṭi*. This word is a combination of the prefix showing diversity (vi) and the word for creation (sṛṣṭi). The prefix *vi* is a short for variety (vividhā) and fascinating (vichitra). Creation is mysterious. The next mantra expresses wonder about diversity in creation.

kō addhā vēda kaḥ ihaḥ pravōcat kutaḥ ājātā kutaḥ iyaṃ visṛiṣtiḥ
ārvāgddēvā asya visarjanēnāthakō vēda yataḥ ābabhūva

Who can claim to know the details about the subsidiary creation (visṛṣṭi)? It is an interplay of a handful of principles which manifested in the previous phase of creation. How and when did this diversity arise? The Devas manifested only during the early part (arvāk) of the diversity phase. They therefore do not know the mystery behind this diversity. Who else can then know from where this mysterious universe arose?

The Nāsadīya sūktam makes a revelation about Devas in this Mantra. Vedic mantras consistently praise Devas as having divine qualities. Devas have power over earthlings. The Nāsadīya sūktam however affirms here that they too do not know the mystery behind creation (visṛṣṭi). This is in contradiction to the common belief that Devas are omniscient.

Consciousness is One. It is the Self, the Brahman. It is the supreme divinity. Its inherent power is undifferentiated energy (svadhā*)*. Undifferentiated energy gave birth to only a handful of formative principles. The germinal seeds of all beings arose from it. Devas are a class of beings. Each Deva rules over a certain set of energies. Devas are different frequencies in the universal energy field. They as energy beings nourish a variety of phenomenon in the macrocosm and in the microcosm. Devas, however, do not know everything about the mystery behind the creation of diversity.

Devas are unlike humans because they do not have a physical body. They have energy or subtle bodies. They shape material existence, which manifests from the *mahimāna* branch of creation. They have a say also in the functioning of the human body.

The universe lacked features per modern science during its very early part. Large scale structures that we see in the universe today formed in a bottom up fashion. The period of 150 to 800 million years from the Big Bang was uneventful. This period is the Dark Age per modern cosmology. Bright objects did not exist during this period. The universe was filled only with invisible Microwave radiation (CMB). The first quasars, early active galaxies and the oldest stars in the universe began appearing at the end of this period. Light appeared in the universe at the end of the Dark Age. The universe moved into exciting times after it was lit up. Atoms began to evolve. The universe added atoms of elements heavier than hydrogen and helium for millions of years. Planetary systems could not host life forms until the universe had generated a supply of carbon and other elements. Several generations of stars had to explode as supernovas to create carbon and other essential atoms. The universe went through eons before birthing creatures of diversity.

Devas appeared per Nāsadīya sūktam in the early part of the diversity phase. We notice that the split in the cosmic egg produced germinal seeds (rētōdhā) of beings. These seeds developed into individual beings later, when the universe was ready. Material transformation was the main activity in the early universe. Mahimāna were intermixing with each other in the early universe.

Material universe is a play of what split from the undifferentiated (svadhā) energy of consciousness. What is the role of consciousness? We only saw a brief mention of it in the third mantra in context of the creation of the universal mind. Did consciousness also diversify during the complexity phase (visṛṣṭi) of creation?

Consciousness operates as individual units of awareness in existence. Sentient beings experience the world because of this awareness. Yoga Vaśiṣṭā compares individual units of awareness to waves and eddies, which arise and disappear in an ocean. Waves appear due to the movement of energy in water. Waves do not create a division in an ocean. They exist as a part of ocean and disappear. Consciousness remains undivided like the ocean. Units of awareness in it are like waves on the surface of ocean.

The Nāsadīya Sūktam completes with a musing about the nature of collective consciousness in the next Mantra.

iyaṃ visṛṣtiḥ yataḥ ābabhuva yadi vā dadhē yadi vā na
asyādhyakṣaḥ paramē vyōman sa anga vēda yadi vā na

This diverse (visṛṣṭi) universe surely arose from something. Does the source sustain the diversity in universe? Is something else sustaining the universe? That mind (anga), the One witness (adhyakṣa) for this creation knows the answer or does it not, while residing in the space beyond (parame vyoman).

The universe came out of Void. What is preventing it from collapsing back? What is holding the universe in place? Is void holding it in place by its power? Physicists have asked these questions. Albert Einstein added a constant to his equation for the general theory of relativity. His equation predicts a collapse if the constant is removed. Einstein named his constant as the cosmological constant. It is the sum of the contribution from the energy of space all over the universe. Thus, the energy of space is keeping the universe from collapsing back! Researchers are working to explain the stability of our universe. Science has some partial answers.

Our Galaxy's ability to retain stars at its outer edge had perplexed astronomers until decades ago. Older gravity models predicted stars to be flying out from the edge of galaxies. Scientists concluded therefore that Galaxies must contain more mass than what has been observed. They explain this anomaly by postulating the existence of "dark matter". Dark matter adds extra mass to a galaxy at its outer edges. Astronomers rely on the idea of dark matter to explain the stability of our own galaxy. The nature of Dark matter however has remained a mystery.

The limit of the actual universe which stretches beyond the limit of the observable universe is unknown. The diameter of the observable universe is fixed. Observations suggest the diameter of the actual universe to be increasing. The actual universe is expanding faster than scientific models predict. Some unknown force is making the universe expand at an accelerated rate. This is Dark energy per scientists. The nature of Dark Energy is a mystery.

Cosmological models of today are credible. They are however based on the idea that 95% of the cosmos is filled with two unknown phenomena, namely, Dark Matter and Dark Energy! Rishis pondered about the stability of the universe eons ago. They came up with different explanations.

The Nāsadīya Sūktam introduces the idea of the One that existed at the time of dissolution in the second Mantra. This Mantra adds a quality to the One. He is the overseer or the witness (adhyakṣa). The One is consciousness or Brahman. Brahman is the observer. The Nāsadīya Sūktam does not call him the creator. The energy of Brahman creates the universe. Brahman remains as the observer. Consciousness remains as a witness while its energy (svadhā) creates the universe. Consciousness is like the sky which is not touched by the clouds. It, as the observer, remains in the space beyond (parame vyoman).

There are two kinds of vacuum energies per theoretical physics. The energy which started the big bang inflation is inflaton. Its density is enormous. Scientists notice however miniscule densities of energy in space. The latter kind of vacuum energy adds to be a big number considering the large expanse of universal space. This substantive number is the cosmological constant in Albert Einstein's equation for general theory of relativity.

One can see a bit of a similarity between Brahman and vacuum. Vacuum with its inflaton energy got the universe started and distanced itself from activities within the created universe! Brahman per Nāsadīya Sūktam did the same. Brahman, however, is filled with pure consciousness. The space of Brahman is untainted like the sky. Brahman, as the observer, remains in the space beyond (parame vyoman). Brahman pervades creation also. We will consider this aspect of Brahman in the next chapter.

Vedic wisdom traditions define three types of space to convey the pervasiveness aspect of Brahman. We will consider the three types of space in Chapter 3. We must understand the principle of mind first in context of consciousness.

The Nāsadīya Sūktam uses an interesting word aṅga in its final line. Is this a reference to the cosmic mind? Aṅga has divisions. The fourth mantra above introduced the genesis of the cosmic mind. The current mantra raises a question. Does the cosmic mind know the answer to the earlier question? What is the difference between the cosmic mind and the One consciousness which existed at the time of dissolution? How is the overseeing aspect (adhyakṣa) of consciousness different from its mind (aṅga) aspect? Yoga Vaśiṣṭā contains succinct answers.

Space (vyoman) carries material creation. Space beyond it is *parame vyoman*. The space which carries our universe appeared inside a tiny nucleation bubble per scientists, and it then began expanding. Scientists speculate about what lies on the other side, at the edge of this space. The space which holds our universe has expanded for 13.8 billion years. A small portion of this with a diameter of 93 Gly is within the purview of science. The observable universe fits within this sphere of space. Galaxies at the edge of this space are steadily moving out and disappearing from the observable universe. Science is facing a challenge with the existing definition of space.

Space of consciousness per Rishis is *parame vyoman*. Material space originated in it. Space which holds the material universe per Yoga

Vaśiṣṭā is *bhūta ākāśa.* Then comes *citta ākāśa* or the mind space. The third space is *cid ākāśa* or causal space. It is indestructible. We will read more about them in the following chapters. The next chapter explores the role of consciousness in existence per Ambhasya Pārē Sūktam.

2. Ambhasya-pārē Sūktam

Creation took place in stages per the Nāsadīya Sūktam. Dissolution is the ground stage. The One existed alone at the time of dissolution. The unmanifested One has no names. Something then arose from the state of dissolution. This is the beginning of the second stage. We do not know the name of this emanation from the Nāsadīya Sūktam.

Salila, or primordial fluid is an important feature of the second stage. It carried the blue print of the universe. A single desire arose during this stage. It became the germinal seed of a Cosmic mind. Cosmic mind carried creation forward to the next stage.

The genesis of the germinal seeds of beings and of the five elemental principles happened after the Cosmic mind woke up. A characteristic of the third stage, per the Nāsadīya Sūktam, is deliberation or effort (prayati). Transformation was inherent and random before this phase of creation. Devas manifested as subtle beings during the early part of the third stage. They hastened creation activities.

The Ambhasya-pārē Sūktam highlights the thread which exists in all stages of creation. This thread is present in today's universe. The Sūktam presents the means to recognize this common thread. This thread has different names in every stage of creation. Some mantras call it Hiraṇyagarbha and others call it Prajāpati and Puruṣa. These names refer to the functions of the unnamed One of Nāsadīya Sūktam. The Ambhasya-pārē Sūktam links ideas in Nāsadīya Sūktam to other Vedic mantras related to creation. It explains the stage of creation when the Devas arose.

Hiraṇyagarbha is the first born, the very first manifestation from the state of dissolution. He carries the Garbha or the womb of creation. This womb contains the primordial soup. The universe lies within this womb in its embryonic state. The name, Hiraṇyagarbha corresponds to the configuration where primordial fluids occupied the universe.

Puruṣa is the name for Consciousness which is within the created universe. The body of the Puruṣa is the manifest universe. Hiraṇyagarbha takes the name Puruṣa after the universe manifests. Puruṣa exists in every being as its consciousness. Puruṣa is as unmanifest as Hiraṇyagarbha and the unnamed entity per Nāsadīya Sūktam.

Hiraṇyagarbha is called Prajāpati, the creator, in the context of the beings which he creates. Prajāpati is the cosmic mind. Nāsadīya Sūktam tells us that a single desire arose within the undivided consciousness. Primordial fluids occupied the whole creation at that time. That single desire became the germinal seed for a cosmic mind. Cosmic mind gave further shape to the universe per its own imagination. Created beings are called *prajā*. Their creator is *Prajāpati.*

The Ambhasya-pārē Sūktam clarifies the idea of a cosmic witness a bit more. The unmanifest One continues to influence every aspect of existence in the created universe. We can say with certainty after studying this Sūktam that the unnamed and unmanifest One of Nāsadīya Sūktam is pure consciousness (Brahman).

ambhasya pārē bhuvanasya madhyē nākasya pṛṣṭē mahatō mahīyān śukrēṇa jyōtigmśi samanupraviśtaḥ prajāpatis carati garbhē antaḥ

The one who is beyond the space of prakṛti (ambha), the one who is amid the world (bhuvana), and the one who is the support of the domain of the Devas (nāka) is subtler than the principle called Mahat. He pervades (sam-anu-praviśtaḥ) the inner faculties (jyōtigmśi) of all beings by his strength (śukra). He as prajāpati, thus moves (carati) within (antaḥ) the womb (garbha) of creation.

The Sūktam begins with the word *Ambha*. Ambha is the ancient most principle of material creation. Nāsadīya Sūktam negated the existence of the mysterious and impenetrable ambha at the time of dissolution. The word ambha here refers to something that cannot be fully occupied. It is the primordial prakṛti. Material space itself originates from prakṛti.

The universe exists in a false vacuum per theoretical physicists. The true vacuum state is different from the false vacuum state. Bubbles of false vacuum nucleate in a field of true vacuum. Many of them die out quickly. One nucleation bubble, however, inflated at the beginning of creation into the objective space, which we see around us. True vacuum continues to exist outside the inflated sphere of false vacuum, the space of our universe. Our universe will dissolve if the space in which we live switched back to its true vacuum state. Ambha, the root of material creation, conveys a similar idea. Ambha is an inexhaustible and undifferentiated field of energy. Ambha is the primordial prakṛti. It is the energy of Brahman.

The next two words of the mantra, namely, *bhuvana* and *nāka* refer to domains in which beings exist. Bhuvana is the material plane of existence. Nāka is the subtle plane of existence in which Devas and other subtle beings exist. These planes are part of creation. Brahman who is beyond Ambha, pervades these two domains.

Vedic texts list existential principles in the order of difficulty of comprehending them. One needs no explanations to understand the idea of a gross body. The difficulty starts when explaining the nature and location of one's Mind. Explanations get increasingly difficult for the ideas of Intellect, Ego and Mahat. The Mahat principle is more difficult to comprehend than the Ego principle. One's individual personality ends with the Ego principle. Beyond that is the *Mahat* principle. Altruism is a characteristic of one whose ego has merged with the Mahat principle. The word *mahātmā* or a great soul is derived from the word *Mahat*. The principle of Mahat can be comprehended to a certain extent because of its association with altruistic people. One cannot comprehend the principle of Brahman by any such examples.

Brahman is more difficult to comprehend than Mahat principle. It pervades the inner faculties (jyōtigmśi), namely mind, intellect, and ego. Human beings are sharp because of the sparkle in their inner faculties. These faculties are anchored in consciousness or Brahman. Consciousness permeates the inner faculties in a delicate and uniform way. The mantra uses the word *sam-anu-pra-vishta*[2] to convey the above idea. Mind, intellect, and ego have no existence without consciousness.

Prajāpati, the cosmic mind, is the creator of the material world. The material world is the thought form in cosmic mind, per Yoga Vaśiṣṭā. The Nāsadīya Sūktam introduced the idea of the origin of a cosmic mind. Cosmic mind and primordial fluids appeared together. Prajāpati, the creator, arose with the arising of the first desire within the primordial fluid. Prajāpati's desire to create began to take a concrete shape with the help of Svadhā, the energy of Brahman. Svadhā split the primordial fluid into three parts. Material creation progressed with the help of the five elemental principles or *mahimāna*. Beings arose from the germinal seeds (rētōdhā), which had also split apart from the primordial fluid. The next mantra acknowledges the role of Svadhā and its connection with the space of Brahman.

yasmin idagm sam ca vicaiti sarvam yasmin dēva adhi viśvē niśēduḥ
tadēva bhūtam tadu bhavyamā idam tad akṣarē paramē vyōman

Undifferentiated energy which Nāsadīya Sūktam calls as svadhā rests in infinite space (vyōman). This space is beyond (paramē) creation and is indivisible (akṣarē). Universe originates from Svadhā and dissolves into it. Devas rest in it. The universe of the past (bhūtam), of the future (bhavyam), and of the present (idam) are all made up of svadhā or undifferentiated energy.

[2] Samanupravishta is a delicately constructed word which is translated in Chapter 3

Svadhā is the cause of both the gross aspect of creation and of beings. *Avyākṛta* or undifferentiated is the adjective for Svadhā. Undivided consciousness is Brahman, and his energy is undifferentiated or avyākṛta. The above Mantra says that this universe arises and merges into avyākṛta. Devas rest (nishedu) in this undifferentiated energy. Brahman's energy stands still during dissolution. It becomes dynamic in creation. Devas realized their oneness with Brahman. They know themselves to be just the energy of Brahman.

Everything that existed (bhūtam) in the past, whatever will be (bhavyam) in the future, and whatever is seen now (idam) are related as an energy field. Energy in this field is undifferentiated, and it is an aspect of Brahman. There is no difference between this energy field and the imperishable (akṣara) and transcendent (parame) space (vyoman).

The universe manifested from an energy field one step at a time during creation. The universe seems to assemble itself "sam eti". The universe merges "vi eti" back into the unmanifest at the time of dissolution.

The first step in creation per science begins within a bubble of false vacuum. A certain energy arises on its own in this vacuum field. It is the "inflaton" energy. It inflates space within the tiny bubble and morphs itself into plasma within the inflated space.

Germinal seeds of beings, including those of Devas, rested once in svadhā, the undifferentiated energy of Brahman. Devas are beings without a gross body. Devas including Virāt, Agni, Indra, and Vayu remain in touch with the energy of Brahman.

Something which is beyond comprehension is called parā or parama. Parā is the adjective in this mantra for the word vyoman. Vyoma is the sky. It is visible to all from everywhere and is transparent. Parame Vyoman per this definition is the deep space which is beyond the sky. Vyoman is unblemished space. It is infinite space which is always still. Parame Vyoman per this definition refers to Brahman. The phrase suggests Brahman to be expansive and unblemished like space. Brahman is beyond space because it is "parame vyoman".

Spacetime did not exist before creation per scientists. Physical space is destructible (kṣara) per Rishis. The only indestructible entity is Brahman which is all pervading like space. It is characterized by the phrase *akṣarē parame vyoman*. Brahman is an incomprehensible (parame) kind of space (vyoman).

Energy which rests in the space of Brahman is the cause for creation. The distinction between Brahman and his energy is subtle but important.

yēnāvṛtam kham divam mahīmca yenādityas tapati tējasā brājasā ca
yam antah samudrē kavayō vayanti yad akṣarē paramē prajāh

Vedic seers (kavi) compare the presence of Brahman in existence (antah samudrē) to threads crisscrossing (vayanti) a piece of cloth. The indivisible One pervades Earth (mahī), Sky (kham), and the subtle domain (diva) of Devas. The Sun (āditya) shines (tapati) bright (tējas) and is radiant (brājas) because of it. Created beings reside in the indestructible (akṣarē) transcendent (paramē) space which is called Brahman.

Brahman is not inert. The surface of the Sun shines with Tējas because of the intensity of Brahman. The rays of the sun have brilliance (brājas) because of the same. The seers compare creation to an ocean (samudra). The ocean is dynamic. Waves arise in it. The internal ocean (antah samudra) is subtle space. Thoughts arise in the mind of beings. These thoughts are waves in the internal ocean. Subtle space is dynamic because Brahman pervades it (viyanti). Created beings (prajāh) are thus a part of the imperishable (akṣarē) and transcendent space.

This mantra emphasizes the proximity of Brahman. Brahman is transcendent space (parame vyoman). Brahman, however, is not far from creation. It is intermeshed with creation. Beings float in an ocean of consciousness like fish in water.

The previous mantra says that the undifferentiated avyākṛta energy rests in Brahman. The first modification of avyākṛta in material creation is manifest space. The Nāsadīya sūktam conveyed this idea as Mahimāna arising from svadhā. Manifest space is one of the five Mahimāna. Space is one of five elemental principles or root elements in creation. Manifest space is 'kham; per this mantra. Brahman pervades kham which is physical space and divam which is subtle space. Brahman's propinquity to beings is assured in gross and mental planes of existence.

yatah prasūtā jagatah prasūtī tōyēna jīvān vycasarja bhūmyām
yad ōśadībhih puruśān paśugsca vivēśa bhutāni carācarāṇi

The Universal genitrix (prasūtī) gave birth (prasūtā) to the world (jagat) from the indivisible One. She created (vycasarja) appropriate bodies for beings (jīvā) in the world from root elements such as water (tōya). The indivisible One or consciousness entered (vivēśa) the bodies of bipeds and quadrupeds in form of their food (ōśadī). The indivisible One thus pervaded material creation (bhutāni) which contains both the mobile and the immobile (carācarāṇi).

The origin of diversity (visṛṣṭi) was a topic of wonder for the seer in the Nāsadīya Sūktam. This mantra discusses the creation of diversity on Earth (bhūmyām). Creation of diversity occurred during a special (viśēśa) phase. Prasūtī, the universal mother of matter caused the universe to be born from Brahman. Five root elements arose from the undifferentiated energy of Brahman. Prasūtī created the bodies for beings from the five root elements.

The universe is called *pra-pancha* because it is a combination or a derivative (pra) of five (pancha) root elements. The root elements intermingled first to create every known phenomenon in the material universe. One or the other root element dominates any material in creation. This has resulted in some materials being solid, some being fluid, some being like air, and some being inflammatory. The world is a complex interplay of materials. Prasūtī creates (vycasarja) bodies for beings (jīvā) from these materials. The word vycasarja (viśēṣēṇa utsṛjat) conveys the meaning that it is a special creation.

The mantra acknowledges the role of water (tōya) in the evolution of life forms on planet earth. The word tōya can be considered also as a reference to the five root elements. Prasūtī assembles material bodies of beings from them. The core of any being, however, is consciousness. The mantra begins with the word yatah (from Brahman) to signify that. Consciousness pervades bodies of lower organisms also. They become food for higher organisms.

Brahman is present in organisms such as ōśadhi. Ōśadhi is a reference to herbal plants, grains, and food material in general. Men and animals feed on ōśadhi. They thus become reinfused with the energy of Brahman. Showers (tōya) nurture ōśadhi. Life forms on Earth, both the mobile and immobile, gain continuous refreshment from Brahman through showers (tōya).

The mantra alludes to the deeper significance of the food chain. Consciousness enters (vivēśa) beings or refreshes the intellect of beings through food. Food sustains life. Food impacts the quality of one's mind and one's consciousness. Māyā controls the functioning of the world in a subtle way through the food chain. Māyā is another name for prasūtī.

atah param nānyad aṇīyasagm hi parāt param yan mahatō mahāntam
tad ēkam avyaktam ananta rūpam viśvam purāṇam tamasah parastāt

There is nothing beyond (parāt param) that (tad) One (ēkam) unmanifest (avyaktam) entity. It is both the minutest (aṇīyasagm) and the subtlest (*mahatō mahāntam*). The unmanifest One is the most ancient (purāṇam) and the indestructible (Ananta) principle. It is untouched by ignorance (tamasah parastāt). The universe (viśvam) is its form (rūpam).

The word param refers to something that is beyond comprehension. The idea of Hiraṇyagarbha is at the limit (param) of human comprehension. Hiraṇyagarbha holds the primordial fluid. Primordial fluids fill the womb (garbha) of creation. What held this womb in place? The Rishis created a metaphor for it. Consciousness holds the womb in place. It is Hiraṇyagarbha. Creation is dazzling like Gold (Hiraṇya).

The Nāsadīya Sūktam avoids naming the One that existed all alone at the time of dissolution. The One is truly beyond human comprehension. The Nāsadīya Sūktam hesitates also in naming the One in which the primordial fluids arose. Primordial fluid, per theoretical physicists, manifested from false vacuum energy. True vacuum continued to exist around a bubble of creation which inflated. There is a similarity between the idea of Rishis and of Scientists. A Rishi's belief differs from only one angle. Consciousness filled everything including vacuum.

The word Hiraṇyagarbha suggests an idea of an early step in creation. Pure consciousness or Brahman is beyond (param) Hiraṇyagarbha. It cannot be understood even with a hint.

Some things in existence are manifest while others are abstract. Manifested entities can be listed as a progression from small to large. Brahman is smaller than the smallest and larger than the largest. Abstract ideas such as Mind, Intellect, and Ego can be listed in an order from the easiest to understand to the most difficult to comprehend. Mahat is a difficult to understand abstraction. It is barely comprehensible. Brahman cannot be understood even to that level. It is an abstraction which is beyond (mahāntam) human comprehension.

There is no division in Brahman. It is one (ēkam). It is beyond the grasp of the senses (avyaktam). The senses derive their power from it. It is permanent. We see it only as the universe (viśvam rūpam). The Vedas accept the antiquity (purāṇam) of Brahman. Rishis did not separate the spiritual from the mundane. Mundane is a part of the spirit. Universe is Brahman.

Beings exhibit different degrees of ignorance. Ignorance is the cause of inertia and dullness. Awareness and dullness are opposite qualities. Brahman is beyond (parastāt) ignorance (tamas). One needs to transcend inertia "tamasah parastāt" to understand Brahman as an experience.

tadēvartam tadu satyamāhuh tadēva brahma paramam kavīnām
iśṭāpūrtam bahudhā jātam jāyamānam viśvam bibharti bhuvanasya
nābhih

The Ambhasya-pāre Sūktam refers for the first time to Brahman by its name in this mantra. Vedic seers (Kavi) consider 'It' to be the Brahman which exists beyond (paramam) existence and as the truth (satyam).

It also has the name ṛtam, the natural rhythm. Vedic rituals such as iśṭā and pūrta are Brahman. Brahman is the axis (nābhih) of the world (bhuvana). Bhuvana had manifested several times before (bahudhā jātam) also.

Vedic seers (kavi) acclaim Brahmā, the creator as the embodiment of the Vedas. Brahman is superior (paramam) to the creator. Rituals prescribed in the Vedas are a part of Brahman also. The above mantra names two representative rituals, namely, iśti[3] and pūrta[4].

The present universe is not the only one of its kind. The mantra makes it clear that universe had arisen many times in the past. Universes in the past existed in Brahman as our present universe does. The only constant entity in the ever changing and evolving world is Brahman. The word Bhuvana refers to a place where beings live. Brahman is the epicenter of Bhuvana. Inert matter and the intelligent have their base in Brahman.

tadēvāgnis tad vāyus tad sūrya tadu candramāh
tadēva śukram amṛtam tad brahma tadāpah sa prajāpatih

Devas such as Agni, Vayu, Sūrya and Candramā are of the nature of Brahman. Deathlessness (amṛta) and shining objects in space (śukram) are Brahman only. Brahman is the Primordial fluid (āpah). In time, It became the cosmic creator (prajāpati).

---

[3] Iśti is a śrauta type ritual which is conducted on full moon and new moon days.
[4] Pūrta is a smārta type ritual. Pūrta is associated with acts such as of digging a pond or a well.

Devas are beings who arose when the universe was ready for diversity (visṛṣṭi). Devas don't know the secrets of visṛṣṭi per the Nāsadīya sūktam. The mantra here lists a few Devas, namely, Agni, Vāyu, Sūrya and Candramā. Hiraṇyagarbha sūktam also mentions these four Devas. Brahman is in all the Devas.

Brahman is indivisible. Devas are different from Brahman only from the viewpoint of duality. Brahman is present before and after every transformation in the universe. Agni, Vāyu, Sūrya, Candramā and other Devas are transformations in Brahman. Brahman is undivided when seen through the eyes of knowledge. The idea of many Devas arises because of the ignorance about Brahman being the sum of existence.

The first transformation in the space of Brahman is the emergence of primordial fluid (āpah). Material existence came out of āpah. Prajāpati, the creator, arose as the cosmic mind within āpah. He molded later phases of creation. The next mantra calls Brahman by the name Puruṣa. Puruṣa's body is the Universe in its entirety. We will read in chapter 7 about Prajāpati and Puruṣa.

sarvē nimēśā jañirē vidyutaḥ puruṣādadhi
kalā murūrtaḥ kāśṭhāscāhōrātrāśca sarvaśaḥ

ardhamāsā masā ṛtavaḥ saṃvatsarasca kalpatām
sa āpaḥ pradughē ubhē imē antarikṣamathō suvaḥ

The idea of Time does not exist at the time of dissolution per the Nāsadīya sūktam. The time principle with its conventional units of measurements is grounded in Puruṣa per the Ambhasya-pāre Sūktam. Puruṣa is a functional name given to Brahman as the indweller in creation.

Time is a subjective principle in Albert Einstein's general theory of relativity. Vedas refined this idea a bit more. Time became measurable only after the advent of Puruṣa. It became a noticeable principle only at the end of the primordial fluid phase of creation.

Every phenomenon in existence has a beginning and an end. They have predictable time spans. Time is responsible for predictability per the Ambhasya-pāre Sūktam. The time span of a phenomenon in nature can fit one of these units of time measures, namely, nimēśā, kalā, murūrta, kāśṭhā, āhōrātrā, ardhamāsā, māsa, ṛtu, and saṃvatsara. The lightning power (Vidyut) of Puruṣa powers time measures. Puruṣa churned (pradughē) the primordial fluids (āpa) to endow Earth, space, and subtle planes of existence with the choicest options.

Puruśā is the cosmic being, the Virāt. His limbs are Devas and his body is the entire universe. Time is a dimension of Puruśā. There were no changes during the period dissolution. Time had merged into Brahman at the time of dissolution. Transformations began when creation gained momentum. Time manifested then as a distinct principle. Rishis created time measures per their affinity to common transformations in nature.

Blinking is an involuntary act. Eyelids close and reopen when one blinks. The time taken for the first half of a blink, namely the time taken by the eyelids to close, is a nimiśā[5]. Nimiśā is the base unit of time. The average duration of a lightning is an order of magnitude less than that. Mantra attributes the origin of the nimiśā unit of time to the instantaneity of lightening which originates in Puruśā.

The other units of time can be derived (kalpatām) from nimiśā. The higher units of time, namely kāśṭhās (~15 Nimiśās), kalā (30 Kāśṭhās), and murūrtaḥ (36 Kalā) are derived from nimiśā. Ahōrātrā is a sidereal day. It consists of 30 murūrtaḥs. Diurnal rhythm is under the influence of the ahōrātrā unit. Fifteen ahōrātrās create an ardhamāsā, and two of them create a māsa. Lunar rhythm is under the influence of the māsa unit. Seasonal rhythms are under the influence of the ṛtavaḥ unit. A cycle of years is under the influence of the saṃvatsara unit. The idea of the above units of time originated (kalpatām) from the intelligence of Puruśā.

Puruśā extracted the finest (pradughē) from the material world (āpaḥ) and endowed the earth with the choicest creations. He similarly endowed space (antarikṣam) and the subtle plane of existence (suvaḥ) with refined creations. This mantra combines the idea of time cycles and the idea of evolution.

Entries in the Periodic table of Elements evolved per astronomers over generations of stars and supernova explosions. Planet earth became ready for life forms over several geological periods. Life forms evolved

---

[5] The average time taken by human eyelids to close is between 0.15 and 0.2 seconds.

over time. The intelligence of the cosmic Puruśā nudged the universe for eons towards progress.

The following mantras focus on the Brahman experience.

nainamūrdhvam na tiryañcaṃ na madhyē pari jagrabhat
na tasyēśē kaścana tasya nāma mahadyaśaḥ

na saṃdṛśē tiṣṭhati rūpam asya na cakṣuṣā paśyati kaścanainaṃ
hṛdā manīṣā manasābhi kḷptō ya yēnam viduramṛtāstē bhavanti

Brahman remains beyond comprehension (pari jagrabhat) because it lacks attributes such as height (ūrdhvam), breadth (tiryañ) and girth (madhya). The renown (nāma) of Brahman who is not controlled (ēśē) by anyone (kaścana) is supreme (mahadyaśaḥ). Brahman is beyond the grasp (saṃdṛśē) of human senses. It has no form (rūpam). No one therefore can see (paśyati) it with their eyes (cakṣuṣā). Those whose mind has transcended agitation (manīṣā) recognize (vidu) its presence in their heart (hṛdā). They reach a state of deathlessness (amṛtā) after thus recognizing (ābhi kḷptō) it mentally (manasā).

Puruśā remains beyond comprehension despite being the dominant contributor in the creation process. He is beyond the purview of sensory perceptions and intellectual understanding. The above two mantras talk about this aspect. One cannot intellectually understand Puruśā in any dimension. The mantra says that neither the vertical (ūrdhvam) nor the longitudinal (tiryam) nor the central (Madhya) axis of Puruśā is known. Puruśā cannot be mathematically measured[6]. No one (kaścana) is an exception to this rule. No one has special access (ēśē).

The sense organs do not experience Puruśā. He does not have color or shape (rūpa) for the senses to distinguish him(saṃdṛśē). One does not see (paśyati) him with their eyes (cakṣuṣā). The wise (manīṣā) recognize his presence in their hearts (hṛdā). They know (vidu) of his existence with a mind (manasā) which is turned inwards (abhikḹptō) and towards the heart. Manīṣā are those that have arrested the flow of their mind toward objects of senses. They have trained their minds through Yoga to rest in the space within.

Manīṣā experience deathlessness. The gross body undergoes death. The wise ones attribute their existence to consciousness within. Consciousness in every being is a part of the immortal Puruśā. Normally, the life force or Prāṇā exits the gross body at the time of death. Prāṇā does not however exit the gross body of a manīṣā at the time of death. It instead merges with the cosmic energy. Yoga texts call this as deathlessness (amṛtatva).

The Ambhasya-pāre Sūktam refers to two other Sūktams at the end of the above two mantras. The first one is the Uttara Nārāyaṇa Sūktam. It describes attributes and the genesis of Puruśā. The second is the Hiraṇyagarbha Sūktam. It talks about the role of Hiraṇyagarbha. We understand the functional distinction between Puruśā and

[6] x-axis (*ūrdhvam*), y-axis (tiryam) and z-axis (madhyam)

Hiraṇyagarbha from these two Sūktams. Puruśā and Hiraṇyagarbha are simply names of cosmic consciousness.

adbhya saṃbhutō hiraṇyagarbha ityaṣṭau

Uttara Nārāyaṇa Sūktam begins with the phrase *adbhya saṃbhutō.* The phrase *hiraṇyagarbha ityaṣṭau* is a reference to the eight mantras of the Hiraṇyagarbha Sūktam. These two Sūktam help our understanding of the following mantras in this chapter. Vedas consider Puruśā and Hiraṇyagarbha to be Devas. Brahman is beyond Vedas also as it is attributeless.

ēśa hi dēvō pradiśōnu sarvā pūrvō hi jātaḥ sa u garbhē antaḥ
sa vijāyamānaḥ sa jaṇiṣyamāṇaḥ pratyaṅmukhāstiṣṭati viśvatō mukhaḥ

viśvatascakṣuruta viśvatōmukhō viśvatō hasta uta viśvataspāt
saṃ bāhubhyām namati saṃ patatrai dyāvā pṛthivī janayan dēva ēkaḥ

The Deva with the name Hiraṇyagarbha is the first (pūrvō) born (jātaḥ). He is omnipresent. He occupies every (sarvā) possible direction (pradiśōnu). He himself (u) is the Brahmāndā, the cosmic egg. Brahānandā is within (antaḥ) the womb (garbhē). He is diversely born (vijāyamānaḥ) as beings. He will be born (jaṇiṣyamāṇaḥ) as future beings also. The difference between Brahman, Hiraṇyagarbha and Puruṣa is syntactic.

Hiraṇyagarbha's face is turned inwards (pratyaṅmukhā). He is aware of divisions in the mental plane and of thoughts flowing in it. He, as Puruṣa, faces outwards (viśvatōmukhō). He sees the events in the universe through the eyes of every created being. Puruṣa is called the universal eye (viśvatascakṣu). The limb of every created being is his own. Puruṣa's legendary hands (hasta) and legs (pāt) have universal (viśvatō) degrees of movement. Hiraṇyagarbha, as Puruṣa, regulates (namati) everything with his limbs (bāhubhyām). He is the Deva who activates the five root elements (patatrai). He shapes (janayan) both the material and the subtle planes (dyāvā pṛthivī).

The second mantra talks about the omni potency of Hiraṇyagarbha. The omnipotent one has taken the shape of all beings in existence. He sees through their eyes and consumes through their mouths. He operates using their hands and is mobile because of their legs. He is called the cosmic eye (viśvatascakṣu), the cosmic mouth (viśvatōmukhō), the cosmic hand (viśvatō hasta) and the cosmic leg (viśvataspāt). He motivates creation to move ahead with his two arms. Dharmā (support for natural order) and Adharmā (contrary to natural order) are his two arms. He created the earth plane (pṛthvī) and the subtle plane (dyu) with the help of the five root elements (patatrai).

There are two causes for creation. Vedic texts illustrate this with the example of a potter. A pot is made of clay. Clay is called the upādhāna or the material cause of a pot. The potter is the nimitta or the instrumental cause of pot. Consciousness or Brahman is the nimitta cause of creation. The root elements are the upādhāna cause. Patatrai refers to the five root elements that have intermingled with each other. The intermingling process is called panchī karaṇa.

The root elements are not in their pristine state in the created universe. Primordial air element dominates the atmosphere. Atmosphere is in a constant motion (air element). Atmosphere however also contains moisture (water element), heat (fire element) and dust (earth element). Primordial earth element dominates molecules and atoms. Solid substances contain chemical bonds (energy element). There are gaps between arrangement of molecules and atoms (space element) within solids. Electrons move constantly across weak bonds among molecules (air element) in fluids and metals. The above are examples of *Patatrai* or the products of intermingling of root elements.

vēnas tatpaśyan viśvā bhuvanāni vidvān yatra viśvam bhavat ēkanīḻam
yasmin idagm saṃ ca vicaikagm sa ōtah prōtasca vibhu prajāsu

pratadvōcē amrutam nu vidvān gamdharvō nāma nihitam guhāsu
trīṇi padā nihitā guhāsu yastadvēda savituh pitāsat

Vēna recognized the universe as resting (nīḻam) in the Oneness principle. He came to this conclusion after observing different spheres of existence (bhuvanāni) in the universe (viśvā). Existence merges (saṃ) into One principle having emerged (vicaikagm) from it. This principle pervades all beings. It is interwoven (ōtah) and inlaid (prōta) in them.

The secret to deathlessness (amrutam nāma) lies (nihitam) deep in the intellect (guhāsu) of beings per Vena, the Gandharva. Three states (trīṇi padā), namely, wakeful, sleep and dream states follow one after another. They are in the cave (guhāsu) of one's own intellect. Vena, the son of Deva Savita understood this and surpassed Savita (savituh pitāsat) in stature.

There is a story about the Gandharva named Vena. These two mantras remind us of it. Vena recognized existence to be just a play of consciousness. He saw everyone go through wakeful, sleep and dream states daily. They recollected dream experienced and analyzed experiences during the waking state and looked forward daily for the sleep state. A very few pondered about what occurs internally during these experiences. He concluded that waking, sleep and dream experiences are a projection of reality from one's own consciousness. He realized that the role of consciousness is not obvious without increased awareness. Awareness is essential to know about the workings of the inner faculties, namely, mind, ego, and intellect.

The intellect can see consciousness to be its foundation. The mind needs to switch into self-referral mode for that to happen. Mind is always in external-referral mode per design. Knowledge about consciousness remains latent until the intellect taps into it. Latent knowledge is the secret which lies deep in a cave of the heart per this Mantra.

We come across several classes of beings in Vedic literature. Many do not have a physical body. Consciousness is at the core also of such beings. Rishis used the metaphor of beings to convey abstractions related to groups and social behaviors. They could refer to a specific abstraction more easily that way. Psychologists today talk about group and social consciousness. Gandharvas are a class of beings who are related to group experiences. The following paragraphs explain an idea related to Gandharvas

Any peak experience leaves a lasting impression on a human being. Stage performers, for example, feel their sway over their audience. Enthralling an audience and receiving accolades is a peak experience. The psyche of a stage performer collects strong impressions after being in an exhilarating lime light. These impressions create a huge aura around famous folk. Rishis considered such aura to be the influence of a Gandharva.

Group psychology is abstract but it effects individuals within a group. There is a temporary bond for example between a performer and his audience. There is an exchange of nervous energy over this bond. Such an exchange impacts the consciousness at both ends. The nervous system of the audience feels an expansion. That of the performer feels exhilaration. Gandharva is the facilitator of this energy movement.

Anything affecting consciousness is conscious per Vedic texts. Gandharva is conscious but not in the same way as a human being. Gandharva consciousness works through the human nervous system but momentarily. It can become entangled in an experience just like a human does. It has the power to become enlightened by becoming aware of entanglements.

Vena is a higher order Gandharva. He is associated with the nervous energy[7] which is directed towards rain bearing clouds. He is a descendent of the Deva by name Savita. Vena hastens the formation of water droplets in a rain cloud. He holds together a cloud region which is ready to precipitate as rain. Clouds whose surface are illumined signal the presence of Vena. The following mantra from Rig Veda hints about the nature of Vena.

ayam vēnascōdayat pṛṣṇigarbhā jyōtirjarāyū rajasō vimanē
imam apām sangamē sūryasya śiśuṃ na viprā matibhī rihanti

Consciousness and its energy make up this existence. Differentiation set in this energy as creation progressed. Rishis recognized the role of differentiation in creation.

Universe per science began as a single energy, namely the Inflaton energy. One energy manifested into subatomic particles and then into atoms, molecules, and complex proteins. The energy of a subatomic particle is difficult to harness when such a particle has become a part of a complex protein molecule. Similarly, an individual being retains limited access to the infinite energy of pure consciousness.

Rishis studied channels in nature which bring cosmic energies to individual beings. Rishis observed the role of rain water in carrying

[7] Humans and animals experience intense longing for showers. They heave a sigh of relief at the sight of showers.

subtle energies to living beings. They recognized Vena to be a part of the rain water channel[8].

Vena believed himself to be a Gandharva before he explored the nature of his existence. He realized that consciousness pervades everything including his own being. He then dropped his limited identity as the offspring of Deva Savita.

The Vena reference in the Ambhasya-pāre Sūktam is a prelude to the idea of ātmā or individualized consciousness. The core of an individual is the ātmā. The Nāsadīya Sūktam talked about germinal seeds of beings. They emanate from Svadhā, the energy of undivided consciousness. Undivided consciousness organized itself into millions of units of consciousness per Nāsadīya Sūktam.

sa nō bandhur janitā sa vidhātā dhāmāni vēda bhuvanāni viśvā
yatra dēvā amrutamānaśānāh tṛtīyē dhāmānyabhyairayanta

paridyāvā pṛthivī yaṃti sadyah pari lōkān pari diśah pari suvah
ṛtasya tantum vitatam vicṛtya tadapaśyat tadabhavat prajāsu

The first born is our creator (janitā) and our relative (bandhu). He is the architect (vidhātā) who knows (veda) the posts (dhāmāni) which an individual being desires and deserves within different domains of existence (bhuvanāni viśvā). Devas know the secret of deathlessness (amrutamānaśānāh). They enjoy (abhyairayanta) their position in the third or the highest domain of existence (tṛtīyē dhāmā).

---

[8] Gandharvas are associated with water at a subtle level.

Brahman (ṛta) pervades (vitatam) existence. Those who contemplate on the emergent aspect (tantum) of Brahman see its influence (tadapaśyat) on all beings (prajāsu). They merge (tadabhavat) with Brahman. Such beings are found all over, namely, on earth and in space and in subtle planes (suvah) of existence. They are within every domain (lōkān) in the universe and in every direction (diśah).

These two mantras repeat an idea which we saw in the Nāsadīya Sūktam. The word "bandhu" appears in the above mantra and in the Nāsadīya Sūktam. A single desire which arose in primordial universe became the germinal seed for the cosmic mind per the Nāsadīya Sūktam. The wise recognize cosmic mind to be the bandhu or the relative. The mantra here begins with the statement "He is our bandhu" and then proceeds to the statement "He is our creator". Is cosmic mind the Creator? We will see a definite answer in Chapter 6 when we study the Hiraṇyagarbha Sūktam. The first born is the creator per that Sūktam. The first differentiation in undivided consciousness signifies the birth of the cosmic mind.

The above mantra calls cosmic mind as vidhātā or the architect. The creator as the architect, distributes skills to individual beings. He is aware of the roles of individual beings including those of the Devas. Roles and privileges are different in different planes (dhāmāni) of existence. Devas enjoy deathlessness in the plane in which they reside. Deathlessness is related to an intellect identifying itself with universal consciousness. Devas are naturally able to merge their individual consciousness with cosmic consciousness.

Other beings can also merge their individual consciousness and can enjoy a special status per the above mantra. Beings who have so succeeded, live in all directions and in all planes of existence. The ability to merge into cosmic consciousness, however, doesn't come effortlessly to beings other than Devas. One needs to contemplate (vicṛtya) on the all-pervading (vitatam) differentiations in the space of undivided consciousness (ṛtasya tantum). One then sees the play of consciousness everywhere. Intellect settles easily with that recognition. The next mantra explains why this contemplation works

parītya lōkān parītya bhūtāni parītya sarvāh pradiśō diśaśca
prajāpatih prathamajā ṛtasyātmanā ātmānam abhisambabhūva

Prajāpati, who was born as the first (prathamajā) from undivided consciousness (ṛta), pervades (parītya) and supports the root elements (bhūtāni). He protects his creation which is spread across many domains (lōkān), in the cardinal (diśa) and other (pradiśa) directions. The lord of beings (Prajāpati) protects them. He instills self-preserving instincts in them from his own being (ātmanā). Prajāpati becomes (abhisambabhūva) the core (atmānam) of beings.

The Nāsadīya Sūktam culminates with the question – Who knows if the source of creation supports the worlds? This mantra affirms that Prajāpati supports everything in creation. He was born from undivided consciousness. He is the cosmic mind. Does mind stuff then hold creation?

sadasaspatiṃ adbhutaṃ priyamindrasya kāmyam
sanim mēdhāṃ ayāsiśam

One's ātmā is the ruler of the dwelling (sadas). He is dear (priyam) to Indra. The splendorous (adbhutaṃ) sadasaspati is most desirable (kāmyam).

The above mantra seeks (ayāsiśam) the fruits of one's action (sanim) and the ability to grasp knowledge (mēdhāṃ) from Sadasaspati. Universe is called Sadas because it rests in the energy of Brahman. Universal ātmā is Sadasaspati. Understanding the truth about the indwelling ātmā is a step in Brahma Vidya, the science of Brahman.

The ātmā connects one to the outside world. Senses and organs of action can function externally because of the energy of ātmā. Organs stop functioning when the ātmā exits the body. The ātmā also connects one to the world within. Freedom and peace follow when mind turns inwards. Turning one's mind inward is a fine skill. This skill develops through self-knowledge. The ātmā is the source of self-knowledge.

The above mantra seeks outward and inward comforts from Sadasaspati. The purpose of human life is internal and external fulfilment.

uddīpyasva jātavēdō'paghnan niṛtim mama
paśūgśca mahyamāvaha jīvanam ca diśōdiśa

The mind has a preferential affinity to the external world even though the ātmā is one's intimate connection to the unmanifest Brahman. A mind's journey towards the ātmā therefore must begin on the outside. A pivot is needed there. Rishis chose Fire to be a pivot for the following reasons.

Fire (Agni) is energy and is an easily relatable representation of Brahman. Fire exists in a subdued way inside the body. It creates warmth. Agni exists as the digestive fire in one's stomach (jaṭara). This Agni is called jātavēda.

Desire shines like a fire. It is the product of Agni within the mind. Jātavēda knows (veda) the fire of desire. The fire of desire burns within the pranic body of an individual being who is born (jātān) with a physical body.

The above mantra expresses a wish to jātavēda to shine bright (uddīpyasva) and to bring prosperity, in the form of domesticated animals (paśū). It seeks Agni's help to chase away nṛṛti, the adversarial devata. The mantra ends with a prayer to Agni to guide one's life (jīvanam) in the proper direction(diśōdiśa).

This mantra reminds us of one more aspect of creation. Everything originates from one consciousness. The energy of Brahman however differentiated during the creation process. Differentiation established new rules and natural laws. Fire is hot, and ice is cold per natural law. One energy of Brahman took on different properties per the will of the creator because of differentiations. Vedas acknowledge this completely.

Rishis contemplated on the nature of differentiations. They discovered that certain materials continue to provide access to different potentials of consciousness. Rishis named them upādhika or the supernumerary forms of Brahman. One can feel the presence of consciousness, for example, in one's Ātmā.

Agni is an upādhika form of the creator. It provides access to the benevolence of the creator. Rishis discovered Vedic rituals to tap into the power of Brahman through upādhika forms of Brahman. The science behind Vedic rituals is vast and mysterious.

mānō higmsī jātavēdō gāmaśvam puruṣam jagat
abibhradagna āgahi śriyā mā paripātaya

The Ambhasya-pārē Sūktam culminates with a prayer seeking the protection of jātavēda. It is a prayer for protection for one's material (jagat) belongings such as cow (gām) and horse (aśvam). The above mantra requests Jātavēda to forgive (abibhradagna) one's mistakes and it invites (āgahi) him to shower (paripātaya) prosperity (śriyā) on the worshipper.

3. *Yoga Vaśiṣṭā and Consciousness*

The Vedic tradition acknowledges three kinds of space. The first is the infinite space of undivided consciousness. The second is the infinite space of divided consciousness. The third is the physical space which hosts the material world. Yoga Vaśiṣṭā introduces these three types of space through stories. It presents ideas about creation against the backdrop of consciousness. Parts of Yoga Vaśiṣṭā appear almost like a discussion on ideas presented in the Nāsadīya sūktam. Yoga Vaśiṣṭā is a part of a larger work by the name Maha-Ramayana. Maharishi vālmiki presents a dialogue in it between prince Rama and his Guru Vaśiṣṭā on mind and consciousness.

Parables abound in Yoga Vaśiṣṭā. They illustrate the truth in parts. Vaśiṣṭā encouraged Rama to grasp applicable parts in parables and to ignore the rest. The text, for example, frequently refers to dream and waking states of consciousness as illustrations. A genuine experience in one of these two turns out to be hollow when viewed from the other. Our concepts about reality are transient in a similar way. A rock-solid concept falls apart upon realizing the pervasiveness of consciousness.

The idea of pervasiveness of consciousness however dodges our normal attention. Guidance is needed to bring one's attention back to it. Our universe does not have a solid existence per quantum physicists also. It is a wave function. This truth also fails to remain in our conscious awareness always.

Vaśiṣṭā frequently compares visible creation to dream objects. An experience of a dream object is real until one wakes up. A return to a state of self-knowledge is like waking up from a dream. Conviction in the existence of the world of forms is a long dream. Wakeful existence per Yoga Vaśiṣṭā begins when one recognizes the connection between consciousness and creation.

The infinite space of undivided consciousness (cid ākāśa) exists in all, inside and outside, as a pure witness. It knows what is real. It also knows what appears to be. It is the substratum for two other kinds of space. The space of divided consciousness (citta ākāśa) creates divisions in time. It pervades beings and is interested in their welfare. Devas orchestrate activities in another type of space while being here. Physical space hosts the root elements, namely, air, water, energy, and matter.

Science is a study of the third kind of space and activities in elemental space. Theoretical physicists recognize alternate states of vacuum energy. These alternate states are not however seen in the space containing our universe. Different kinds of space per Vedic texts is like this.

Nothingness is the common factor among physical and other two kinds of space per Yoga Vaśiṣṭā. Mind space (citta ākāśa) is empty. One can compare it to the inert nothingness of physical space (bhūta ākāśa). Thoughts are but energy impulses. They appear and disappear in mind space. Matter in physical space similarly assembles and disperses. It forms different configurations in space. The space of pure consciousness (cidākāśa), is more like mind space. It however remains empty all the time.

Experiences in the world of matter sustains mind space. The space of pure consciousness however remains unaffected by happenings in mind space. The space of pure consciousness per Rishis is Brahman, truth, god, Shiva, void, and supreme self.

Cosmic consciousness experiences its own creation through trillions of individual beings who are localized sparks of consciousness. A unit of consciousness (ātmā) resides in the heart of every being. This individual unit of consciousness forgets its Brahman nature. It thus contracts into finite intelligence. This intelligence holds on to a conviction that it is body bound. It rests in mind space as the intellect associated with a body. This intellect accepts and rejects ideas and experiences per its core conviction. The ability to remember its true nature fades to the background and it becomes dormant. Finite intelligence becomes bound to the space of divided consciousness. It can regain its freedom by recognizing mind space to be sustained by fictitious movement of energy in consciousness.

We review the Nāsadīya sūktam in this chapter in the context of Yoga Vaśiṣṭā. We are thus able to gain new insights into the meaning of each mantra of the Nāsadīya sūktam

ṇasadasinnō sadāsīttadānīm nasīdrajō nō vyōmā parō yat
Kimāvarivaḥ kuha kasya śarmannambhaḥ kiṃāsīd gahanaṃ gabhīram

Consciousness is the cause for dissolution of material existence per Yoga Vaśiṣṭā. Consciousness created the universe by entertaining a core belief "this is real". The third mantra in the Nāsadīya sūktam corroborates this idea. Mindspace originated with one desire. Consciousness experiences creation only through Mindspace. Consciousness can similarly dissolve Mindspace. It only needs to entertain a reverse desire. Physical space dissolves because of the dissolution of Mindspace per Yoga Vaśiṣṭā. The parable of the dance of Rudra steps us through the dissolution of physical space and mental space. It is presented towards the end of this chapter.

The pair of opposites, namely real (sat) and unreal (asat) occurs in the above mantra. They are complementary per Yoga Vaśiṣṭā. An object experienced in a dream is both real and unreal. Dream objects are real while they are being experienced. They become unreal when a dream ends. Mind is real when it experienced as real. Mind is void when it is without thoughts. Mind is therefore both real and unreal. It therefore did not exist at the time of dissolution. All inner faculties are like the mind. They are both real and unreal and were therefore also absent during dissolution.

Human body is real and unreal. The notion of 'I am this body' is based on a conviction. One equates one's self to a mass of flesh and bones per this conviction. This conviction arises because of forgetting one's true nature. We find a similar idea in modern science. Matter has no solidity from the angle of the quantum theory. World is real when we ignore quantum viewpoint. World is unreal otherwise. The world which is both real and the unreal had ceased to exist in the state of dissolution per the above mantra.

ṇa mṛtyurāsīt amṛtam tarhi na rātriyā anha asīt prakētaṃ

ānīt avātaṃ svadayā tadēkaṃ tasmāt ha anyanna para kiñcanāsa

The above mantra refers to the One who existed during dissolution and to his energy *svadhā*. Vedas call the One as the supreme Lord, the eternal, the unborn, the self-effulgent, the Ātmā, and the one beyond description. Vaśiṣṭā was confident about what continued to exist during dissolution. It was pure consciousness. Dissolution is a state of perfect equilibrium. Subject-object relationships had ceased in this state. Objective universe had merged completely into cosmic intelligence.

A fiery circle is created by whirling a ball of flame. Whirling movement sustains the circle. The circle vanishes when the ball is held steady. World fits this analogy when we see infinite consciousness with its associated energies to be a ball of flame. World appears when there is a movement in energies. World disappears when that movement stops. Consciousness alone exists when all vibrations in energy stop. The energy of consciousness is *mahāspanda* per Yoga Vaśiṣṭā. The word *spanda* refers to the act of flashing suddenly. Mahāspanda is but svadhā of the Nāsadīya sūktam. We can interpret the sentence – "The One inhaled with the power of svadhā" considering Yoga Vaśiṣṭā. The One flashed into existence with the help of Mahāspanda.

The distinction between Brahman, the One, and Mahāspanda is purely verbal per Yoga Vaśiṣṭā. Vibration began in state of equilibrium without a cause. It was a natural happening. Infinite consciousness then becomes aware of this energy movement, even as a person becomes suddenly aware that he has been day dreaming. This awareness is the seed for subject-object relationship. It is a vital part of cosmic order. Cosmic order is *Niyati* per Yoga Vaśiṣṭā. The distinction between Brahman and Niyati is also verbal. The distinction between real and unreal becomes smudged when subject and object relationship becomes an enduring impression. *Māyā* is the inherent power of Brahman which causes enduring impressions. The distinction between Brahman and Māyā is verbal.

Citṣakti is another name for the energy of infinite consciousness. It is in motion all the time. Citṣakti dictates the nature of every object in the universe. Everything in existence gains its characteristic quality because of Citṣakti. The fifth mantra in the Nāsadīya sūktam corroborates this idea.

Mighty forces and the germinal seeds of beings originated from svadhā per that mantra. Svadhā created the distinction between mighty forces and germinal seeds of beings. Another name for Svadhā is Citṣakti. The five root elements gained their specific qualities per the rule of Citṣakti.

Infinite consciousness and its energy appear as though independent of each other because energy is always in motion in the universe. Vaśiṣṭā and other Rishis however had experienced the merging of Citṣakti and the supreme Self. Knowledge, self-will, and dynamism are integral to the supreme Self. They however appear to be different from consciousness after manifesting.

tama āsīt ṭamasā guḻhamagrē aprakētaṃ salilamā sarvagm idaṃ
tuccyēna abhvapihitaṃ yadāsīt tapasas tan mahinā jāyataikaṃ

Dream objects are projections from one's consciousness. One's intelligence is clouded during a dream. The mind on that account believes dream objects to be real. Ignorance dominates dream state. The relationship of creation to Brahman is that of dream objects to a mind. Brahman projects the appearance of creation from its own consciousness. Ignorance appeared at the beginning of creation. A bit of ignorance was needed to make an unreal dream like object appears as real per Vaśiṣṭā's instructions to Rama. *Avidya* or ignorance is an energy of consciousness.

The Nāsadīya sūktam refers to the principle of inertia or ignorance (tamas) in the above mantra. An excess of this causes sleep. A moderate amount of this dulls the intellect in the dream state. Tamas is a quality of *prakṛti*. Prakṛti has three Gunas or qualities, namely Sattva, Rajas and Tamas. Prakṛti is another name for the energy of consciousness.

The infinite effulgent consciousness created the knowable within itself. The first step of manifestation during creation was the emergence of a cosmic egg. The cosmic egg became a knowable entity within the space of consciousness. Space of consciousness refers only to the infinite nature of consciousness. Prakṛti, the energy of consciousness, emerged per Yoga Vaśiṣṭā at the onset of creation. A cosmic soul emerged only after a considerable time when prakṛti energy had gained strength. The Nāsadīya sūktam calls prakṛti by the name svadhā.

Mantras consider cosmic soul to be the first conscious entity in creation. The first entity was *Hiranyagarbha*. *Hiranyagarbha* carried the cosmic egg. Salila, the primordial fluid mentioned in the Nāsadīya sūktam lay within this egg. Infinite consciousness seemingly limited itself as a *jīvā* or a being at that point per Yoga Vaśiṣṭā. We will read about the idea of jīvā in a bit. The universe in the form of a cosmic egg became ready for a cosmic mind. The transformation of the infinite consciousness into Hiranyagarbha was not real.

kāmastadagrē āmavartatādhi manasō rētaḥ pratamaṃ yadāsīt

satōbandhumasati niravindan hṛidi pratīśyā kavayō maniṣā

The above mantra says that a desire arose at that point. It then became the germinal seed of a cosmic mind. An intention to create the universe was the first desire. There are two assumptions behind the above statements. The first assumption is that the entity which experienced the desire was conscious. The second is that the entity considered itself separate from infinite consciousness. I-ness per Yoga Vaśiṣṭā is ego-sense. The dissolution of this ego-sense causes dissolution. We will see a description of this towards the end of this chapter in the section on the dance of Rudra.

The cosmic being is *Virāt, Svayambhu, Brahma* per Yoga Vaśiṣṭā. Consciousness of Brahma became aware of energy moments all over. There was no 'externalized' awareness before these energy movements. Brahma sensed primordial fluid everywhere. The fluid was in vigorous flux because of energy in it. That energy however was only an integral element of pure consciousness. Brahma considered the energy movement to be his own life force because his awareness could not exist without being aware of an 'other'. *Prāṇa* or life force of Brahma is Citśakti, the energy of consciousness.

Brahma has no mind as he himself is the cosmic mind. Creation is within Brahma as dream objects are within a dreamer. Brahma's creation is made up of consciousness and its energy. Movement of energy always accompanies Consciousness. It subsides only at the time of dissolution. Creation merges back into pure consciousness at that time. Everything in the universe exists as Brahma's dream until then. The universe has gained its firmness because of the persistent nature of his dream. Vaśiṣṭā narrated a story to reiterate the above points to Rama. Brahma witnessed others like him creating their respective universes in that story.

Beings who arose later in infinite consciousness after Brahma were also in Brahma's dream. They were subject to the rules of Brahma's mental creation though they were made up of pure consciousness. They could entertain their own desires while being limited by the rules within Brahma's creation. Infinite consciousness alone experienced all the pleasures and the tainting from them. Sages like Vaśiṣṭā arose in the beginning within Brahma's dream. Brahma wanted a few enlightened beings in his creation. Such beings guided others back to their infinite nature. Vaśiṣṭā narrated the following account about himself.

Creator brought me into being. He drew me to himself and drew the veil of ignorance over my heart. Instantly, I forgot my identity and my self-nature. I was miserable. I begged of Brahma the creator, my own father, to show me the way out of this misery. Sunk in my misery I was unable and unwilling to do anything, and I remained lazy and inactive. In response to my prayer, my father revealed to me the true knowledge. That knowledge instantly dispelled the veil of ignorance that he himself had spread over me. The Creator then said to me: "My son, I veiled the knowledge and revealed it so you may experience its glory; for only then can you understand the travails of ignorant beings."

Vaśiṣṭā refers to the heart in his narration above. Nāsadīya sūktam also corroborates the heart connection. One's heart is the seat of Prāṇa in the human body. A false conviction about one's nature arises from this region because a tiny bit of ignorance veils knowledge there. Heart is also the seat of desires. The false conviction taints one's desires. A desire creates thoughts in the mind. Thoughts in turn create other desires.

Yogis train to make their Prāṇa stand still. Mind comes to a standstill by it. Mind then merges with the heart, the seat of feelings. This allows the intellect to pierce through the veil of false conviction. The Nāsadīya Sūktaṁ refers to this sequence in the mantra starting with the statement – 'one desire arose and it became the seed for the cosmic mind. '

Mind gets confused when presented with multiple ideas at one time. Diversity in universe is beyond its comprehension. Heart accepts diversity easily per Vaśiṣṭā's teaching to Rama. Living beings conjure complex realities within the heart. Such illusions gain strength through individual life experiences. The illusion breaks when a one questions basic assumptions about creation. Intelligence when directed into one's heart is therefore beneficial. It removes doubts about the nature of one's consciousness. This purifies the mind. The conviction that all of one's experiences are within the dream-creation of Brahma then gains a strong foothold.

Infinite consciousness shines in one's heart when such opposites as attraction and repulsion, love and hate have ceased. One is otherwise stuck to a false idea of existence. An object exists only within a subject. This per Vaśiṣṭā is like an ornament existing potentially in gold. Yogis and sages recognized the origin of the human mind because of their expanded vision. The Nāsadīya sūktam says that the sages searched for the truth and found it in their heart. The link between inert body and intelligence can be found in one's heart. This link is ancient per Sages.

The link appeared when the seed for the cosmic mind emerged in creation per the Nāsadīya sūktam.

Infinite consciousness resides within the space of one's heart as a spark. Intellect in a human body entertains a severely limiting belief even though it was born of this spark. It believes itself to be the body which is born of parents. Spiritual practices substitute this belief with an alternate conviction. Yoga Vaśiṣṭā lists three appropriate alternatives. They are (a) I am the soul which is in this body, (b) my soul is all pervading, and (c) I am pure void like space.

Desires arise despite one being on the spiritual path. Desires however do not create bondage when they are concomitant to one of the above three alternate convictions. Vaśiṣṭā's first preference is the last of the above three convictions. Vaśiṣṭā narrated several stories to Rama to affirm his preference. Vaśiṣṭā guided Rama to engage in Self enquiry while holding the conviction that the Self is void like space. The Nāsadīya sūktam refers to the importance of Self enquiry with a single word, namely, *pratīśyā*.

tirascīnō vitatō raśmirēśām adasvidāsīt uparisvidāsīt
rētōdā āsan mahimāna āsan svadhā avastāt prayati purastāt

The Nāsadīya sūktam introduces the idea of the creation of the manifest universe in this mantra. Vaśiṣṭā expands on the same idea. Brahma desired to experience sound and physical space manifested in support of it. He intended to experience touch and air element manifested within space. Brahma could not see space and air. They were too subtle to be seen. He desired to see his creation. Fire element manifested as a result. Fire elements is the power behind every source of light in the universe. Brahma desired for a complement to the heat of fire. Water element manifested. It brought taste experience along. Earth element arose when Brahma wished to smell his creation. Brahma continued to be a subtle being even after the birth of these five root elements. He remained as the mind of the universe.

The Nāsadīya sūktam refers to the above five root elements with one word, namely *mahimāna*. Mahimāna arose from svadhā per the Nāsadīya sūktam. Svadhā is the energy of consciousness. Brahma, the mind of the universe is of the nature of consciousness. Root elements are therefore a play of the cosmic mind and the energy of consciousness. The energy of consciousness has many names. As prakṛti that energy birthed the five root elements. The root elements are as void as pure consciousness per Yoga Vaśiṣṭā. The entire universe is made up of one energy per modern science also. It is just a wave function.

Mahimāna created the inert branch of Brahma's creation. Consciousness is equally present in the inert branch of creation per Yoga Vaśiṣṭā. This branch abandoned its infinite nature at the time of creation. Consciousness elsewhere began perceiving itself as infinite sparks in space. The Nāsadīya sūktam conveys this same idea. The seeds of beings arose along with mahimāna. A spark of consciousness is the core of every species in nature while the root elements constitute its body. Prakṛti, the energy of Brahman was responsible for shaping those bodies.

Māyā, injected an element of ignorance into creation. Māyā is another name for the energy of consciousness. Ignorance mentioned here is of the nature of self-forgetfulness. Consciousness surrendered to ignorance while respecting the power of its own energy! Vaśiṣṭā expounds on this theme a bit more

Consciousness began thinking of itself as individual sparks. Each spark entertained an ego-sense. Individual sparks of consciousness are *Rētōdhā*, the germinal seeds of beings, per the Nāsadīya Sūktaṁ. Yoga Vaśiṣṭā explains the nature of the seeds of beings. Each spark of consciousness along with an ego-sense is the *jīvā*. A jīvā has the potency of Brahma because it is made up of consciousness as Brahma is. It conceives of a body for itself. Rules of Brahma's creation however apply to this body. A jīvā's mental conception results in a physical body for itself. Physical bodies are made up of root elements per Brahma's rule for this creation. A jīvā, at that point is confused by *Avidya* which is associated with material creation. Jīvā then forgets its original nature. Avidya is another name for Māyā. Jīvā in its fetal sleep entertains the conviction that it is physical matter. Thus begins the journey of an individual spark of consciousness into a lifetime of limited intelligence.

A strong sense of "I" or *ahaṃkārā* which is tied to the physical body is the seed for psychological conditioning. Individual beings can experience their oneness with cosmic consciousness despite this. They must undertake a reverse journey for that. The reverse journey pierces past Avidya. Avidya or ignorance, in this connection, is a lack of knowledge about the fictitious nature of Ego-sense.

Vaśiṣṭā conveyed the illusory nature of ego-sense to Rama through several parables. He made the following points through those parables. Ego-sense objectifies experiences. It forgets that no experience is possible without consciousness. Judging faculty which accompanies ego-sense gains strength by forming concepts about objects. *Buddhi* is the name for the Judging faculty in humans. Buddhi forms concepts and notions based on experiences. It however is only an adjunct to ego-sense. Buddhi itself has an adjunct whose name is *manas* or mind.

Thoughts hover around concepts created in one's Buddhi. Sense experiences generate thoughts also. Thoughts form the mind. Buddhi is the seed of the mind or manas. Buddhi becomes vibrant because of manas. It adds strength to ahaṃkārā. Ahaṃkārā, Buddhi and Manas make up one's subtle body. Subtle body is also called as the *ativāhika*. It is more robust than one's gross body per Vaśiṣṭā. Gross body perishes at death while the ativāhika proceeds to the next birth. The latter is of the nature of energy. Energy is not destroyed per laws of thermodynamics.

Ativāhika body has a cosmic connection. An ativāhika body is just thought forms in the cosmic mind. It exists in Mindspace. An idea about one's next gross body gains strength in the ativāhika body at the time of one's death. It becomes a reality because of the strength of conviction there. Mindspace thus controls events in physical space. Forces in physical space help in creating a physical body according to a notion in Mindspace. Yoga Vaśiṣṭā highlights the longevity of the ativāhika body through several stories. Most of such stories are about the cycle of rebirths. Beings arose at the time of creation. They have cycled through diverse forms. They will be reborn based on mental conditioning at the time of death.

Creation was not hierarchical beyond a certain point per Yoga Vaśiṣṭā. Mindspace is a space of divided consciousness because of ativāhika bodies. Each Ativāhika body creates a reality for itself. The universe of diversity is chaotic. The following mantra reflects this theme *kō addhā aēda kaḥ ihaḥ pravōcat kutaḥ ājātā kutaḥ iyaṃ visṛiṣtiḥ ārvāgddēvā asya visarjanēnāthakō vēda yataḥ ābabhūva*

Diversity is a hallmark of the universe. It is surprising that all this began from a single entity. That entity split initially into only two principles, namely the mighty forces (Mahimāna) and the germinal seeds of beings(rētōdhā). Mahimāna are only five in number. It is hard to believe that such few principles exploded into such diversity. Devas also do not know the secret behind the proliferation of diversity and the mysteries in nature.

Devas are beings who have only a subtle body. They have a mind and an ego-sense. All groups of beings including Devas, Asuras, Gandharvas and Yakshas have a mind per Yoga Vaśiṣṭā. Different classes of beings exhibit different level of Avidya or ignorance. Devas entertain a tiny bit of ignorance. Trees and shrubs exhibit deluding levels of ignorance. Insects and worms live with blinding levels of ignorance. Cardinal devas, a category of devas, became enlightened early on. They preside over the universe as helpers of Brahma.

Sages realized that they are a part of cosmic consciousness. Enlightened beings like them play a part in the universe with full awareness. Beings of every class can reach enlightenment per Vaśiṣṭā. Vaśiṣṭā narrated stories of Devas, Asuras, Kings, robbers and even a crow which became enlightened. Enquiring into the nature of the mind is the first step towards enlightenment.

Creation is not a real transformation but is merely a creative-thought inherent in the cosmic being per Vaśiṣṭā. Five root elements which themselves are only ideas in the cosmic mind combined with each other in different proportions. The products of that combination filled the physical space. The combination of root elements is the reason for the variety in materials and phenomena. The intricately formed human body disintegrates into five root elements, namely, earth, water, heat, air, and space when consciousness departs from it.

Every being in the universe is a spark of consciousness. Imagination flourishes in individual minds. Creativity floods the universe after arising from the intelligence of every class of being in the universe. Creation takes place in the universe in every corner in the space of pure consciousness. Creativity is the very nature of consciousness. Devas are bound to the confines of one Brahma's creation. They can therefore not know about other spheres of existence.

The space of pure consciousness remains undivided. The energy of pure consciousness is forever in motion. Brahma arose as the cosmic mind in this energy movement. He imagined creation within it. Brahma's imaginations match the configurations of energy movements and vice versa. This per Yoga Vaśiṣṭā is identical to the projection of dream objects in the consciousness of a sleeping person. The space of Brahma's creation is also infinite but is filled with divided consciousness. Devas are compartments in this space of divided consciousness. They help Brahma in molding the elemental space of creation. Devas have purview only to parts of creation. The Nāsadīya Sūktam supports this idea. Devas are not capable of knowing creation completely because of when they originated.

Brahma exists as both pure consciousness and as thoughts. Thought is always subject to confusion per Vaśiṣṭā. Brahma, though unborn appears to come into being. He supports his creation, the entire universe with his intelligence. An analogy for this is the collapse of a dream object upon the dreamer's intelligence dropping a corresponding thought. Every thought which arises in Brahma's intelligence creates a form.

Forms exist in physical space. There are however corresponding thoughts in Mindspace. Forms appear robust because of the Self-forgetfulness of the truth about thoughts and forms per Vaśiṣṭā. Goblins are formless and are yet seen to have forms because of a perceiver's delusion.

The Creator is exempt from self-forgetfulness per Vaśiṣṭā. He never imagined a body for himself. He is therefore never in the material plane like other beings. Vaśiṣṭā explains the riddle of the creation as follows.

Creativity is inherent in the cosmic being. Early creation occurred purely based on his creative thoughts. They were not real transformations. The faculty of sound for example manifested itself in space when Brahma wished for it. The principle of Ego and the principle of time arose simultaneously in a similar fashion. The ego principle was a vital element for subsequent creation. Brahma narrated the following to Vaśiṣṭā. It illustrates the spontaneity of early creation

My child, listen to what happened to me at the beginning of this epoch. Cosmic night followed the end of the previous epoch. I woke up at the end of that night. I first completed my morning prayers. I then looked around. I could see the infinite void all around. The void appeared to me as neither dark not illuminated. I desired to create and immediately began to see subtle visions in my heart. Several seemingly independent universes were present in those visions. My mind alone saw those universes. My child, it is only the mind that appears as all this.

Cosmic consciousness alone perceived the universe of diversity within itself even as the dreamer perceives diversity within himself. The dreamer sees diverse objects within the inner void.

Jīvā's case is only somewhat different from the experience of the creator. jīvā perceives its body to be made up of the elements in the void. Brahma however doesn't have a body because he never perceived one. A Jīvā's psychological conditioning begins with a notion of a physical body. It deepens with every birth. New notions and experiences of all kinds increase a Jīvā's psychological conditioning with every new birth. That conditioning wanes when curiosity arises about reality. Ego drops when one pursues self-knowledge seriously. A sense of oneness remains at the end. That is liberation per Vaśiṣṭā.

A desire leaves behind an impression till one is liberated. Cravings and aversions are the nature of those impressions. A contact with an external object fans more desires when such an impression is present. Desires are generally binding in nature. Desire arises also in the mind of a liberated person. It however is not associated with a craving. Such desires are per one's natural functions. They are not binding. Creator for example desired to create even before encountering objects. Such desires arise naturally in enlightened beings. They are not binding. A liberated one's attention moves outwards naturally without ego. The feeling 'I want this to be mine' does not arise in them.

Thought form conceives and perceives other thought forms. This is strange but true. Existence is a continuous experience of thought forms.

The stream of thought forms begins with ego sense. Ego sense arose early in creation. Consciousness conceived of an idea within itself at that point. It believed that it had perceived the idea as an object. Brahma arose in this way. Countless sparks of consciousness arose along with the root elements from primordial fluid per the Nāsadīya Sūktaṁ. Ego sense arose in those sparks of consciousness.

Vaśiṣṭā illustrated this through several stories to Rama. Those stories illustrate an instant appearing to be an epoch or multiple epochs or divisions of epochs. Vaśiṣṭā also narrated stories to illustrate thought forms being not restricted to sentient beings alone. Cosmic consciousness creates thought-forms everywhere. Consciousness exists in inanimate objects as well. The drama of self-veiling and self-knowledge can enact in a rock and inside an atom.

There is nothing within and outside Brahman. An observer is of the nature of Brahman. The observer is an infinitesimal part of Brahman and imagines world-appearance there. Brahman remains empty despite the imagination of an observer because world appearance is only imaginary. A feeling of "I", sustains the world appearance in an observer's corner of the space of consciousness. The feeling of "I" itself is the observer who is an infinitely small portion of Brahman. The feeling of "I" cannot divide the infinite space of Brahman because it is unreal. It creates a division in Mindspace. Mindspace arises from the undivided space of consciousness.

Yoga Vaśiṣṭā provides a deeper analysis of the creation process than the one which we see in the Nāsadīya Sūktaṁ. It redefines often used terms which explain the process of creation. It uses the phrase 'in the beginning' only for instruction purposes. It does not refer to an actuality.

A notion of Creation appeared spontaneously in one corner of the space of undivided consciousness. The notion of duality accompanied it. Creation considered itself separate from the field of undivided consciousness per that notion of duality. The idea of a separate existence caused imaginary divisions and differentiations in the otherwise pure consciousness.

Consciousness considered itself a living entity or *jīvā* because of an idea of a separate existence. There is an 'other' in an idea of separateness. Consciousness had to become aware of the 'other' and therefore Vaśiṣṭā calls it "cit" or awareness. Consciousness began looking for distinguishing marks among 'objects' and got the name *Buddhi* or the faculty of judgement. Consciousness began making concepts and percepts related to those objects and got the name *manas* or mind. The mind developed a proclivity to focus on the world of objects. That proclivity hardened within consciousness into a conviction of 'I do ...'. Consciousness thereby got the name Ahamkara or Egotism because it entertained the idea of a 'doer'. A weak sense of "I am" is ego sense per Vaśiṣṭā and it is different from Ahamkara or egotism. The field of consciousness which hosts the named aspects of consciousness is citta or psyche per Vaśiṣṭā.

Billions of sparks arose in consciousness at the beginning of creation. Each spark entertained a notion of a separate existence inside the space of undivided consciousness. That notion ended only when an individual being attained enlightenment. The idea of a separate existence got the name Avidya because it disappears.

Avidya or ignorance is a rudimentary quality in creation. It is a quality of prakṛti. Prakṛti is svadhā per the nāsadīya sūktam. The nature of the seed of an individual being is ignorance per Yoga Vaśiṣṭā. An ativāhika or a subtle body develops around the seed of ignorance. It eventually gains a physical body. Nāsadīya sūktam mentions only the creation of the seeds of beings. Germinal seeds emerged from svadhā, the primordial fluid along with the root elements.

Material creation arose from mahimāna, the root elements of creation. Mahimāna arose from the undifferentiated energy of consciousness or svadhā. The root elements are a modification of the invisible primordial energy. Vaśiṣṭā therefore calls material world to be an illusion. The gross body of an individual being is a different kind of a void. The subtle body upholds the notion of a gross body.

A subtle body reflects the universe like a mirror reflects an object placed in front of it. An object inside a mirror is imaginary. The cosmic being has a subtle or ativāhika body per Yoga Vaśiṣṭā. That being imagined a new subtle body for itself at the end of the last dissolution. That subtle body identified itself as Brahma. Other jīvās accepted diversity in creation as a truth because Brahma imagined this diversity. The vision of multitudes of individual beings is an illusion as explained by Vaśiṣṭā.

Consciousness plays as trillions of jīvās in this creation. A jīvā rides on its ego sense. Ego sense rides on intelligence, intelligence on mind, and mind on Prāṇa. Senses feed Prāṇa and the body hosts the senses. Evolution has upheld this dependency chain. Body appears to be supporting all others in this chain. Prāṇa is a vital link in this chain of dependency. The link between mind and the senses needs Prāṇa.

Mind normally moves to the rhythm of Prāṇa. Prāṇa is unable however to move a mind which has merged with the heart. Conversely, the mind attains a quiescent state when Prāṇa becomes steady. Yogic practices utilize these two principles. They free consciousness from the clutches of material existence. Consciousness then reverts to its witness state. The following mantra refers to the witness state of consciousness

iyaṃ visṛṣtiḥ yataḥ ābabhuva yadi vā dadhē yadi vā na
asyādhyakṣaḥ paramē vyōman sa anga vēda yadi vā na

Infinite forms arise in material space. Intellect which resides in one's subtle body associates a name to each form. Subtle bodies exist in the space of divided consciousness. Infinite names exist only in the space of divided consciousness. The space of divided consciousness is at the root of the idea of a diverse creation.

Rama asked Vaśiṣṭā what keeps the universe in place? Diversity in creation is imaginary and it needs no support explained Vaśiṣṭā. There is also an alternate explanation. The Nāsadīya sūktam conveys the idea that the intellect which is upholding the idea of diversity is unsubstantiated.

The knower of the field of forms and names is the witness aspect of consciousness. Human body is called the *kṣētra* or the field. Witness consciousness is present in the heart of a human. It is called *kṣētrajña*, the knower of the field. Kṣētrajña knows all facets of the field of the body and its vicinity. Ego sense exists only as an appendage to witness consciousness. Ego-sense becomes entangled in the field. Witness consciousness remains untouched.

The Nāsadīya sūktam refers here to two aspects of consciousness. One is the witness (adhyakṣa). The other is the mind (aṅga). The mind entertains ideas of divisions and is conditioned. It normally does not notice the existence of witness consciousness within. Witness consciousness lies hidden in all beings. The mind becomes aware of the witness consciousness only when it drops its conditioning. The boundary between an individual unit of consciousness and universal consciousness disappears then.

Witness consciousness pervades all three states of normal consciousness, namely, waking, dream and deep sleep. These three states of consciousness are hidden in the cave of one's heart per the Ambhasya pārē sūktam. Witness consciousness transcends all normal states of consciousness. Sages call it the turīya or the fourth state of consciousness. Liberated sages remain in it constantly. Others however slip in and out of it without awareness.

Thought is the trademark of active and dream states. Inertia is the hallmark of sleep state. Absence of ego-sense is the seal of Turīya state per Vaśiṣṭā. A jīvā cannot exist without ego sense. It must transform into cosmic consciousness without ego sense.

Witness consciousness rests per the Nāsadīya sūktam in its home, the space beyond (parame vyoman). Divisions and diversity abound in Mindspace but not in the space of undivided consciousness which is the space beyond. When ego sense merges into that space, memories and concepts which were associated with its ativāhika body fade away. Mind which has merged with witness consciousness does not see divisions.

There is no division from a vantage point in Parame Vyoman. Parame Vyoman carries Mindspace. The last two sentences appear to contradict each other. The Nāsadīya sūktam highlights this contradiction in the form of a question – Does witness consciousness (adhyakṣa) know the answer? Can it as the mind (aṅga) answer the question about diversity in creation?

Universe collapses when the ego sense of creator dissolves per Yoga Vaśiṣṭā. The ego sense of the cosmic mind upholds the idea of a universe. The ego sense of the creator is however a notion and is therefore a void. Vaśiṣṭā illuminated the above idea with the following parody to Rama.

A universe began dissolving in one corner in the space of undivided consciousness. The ego-sense of its creator began to dissolve also. This shook up the universal order. Material objects began disintegrating into the five root elements which made them. This disintegration was chaotic.

Vaśiṣṭā perceived the universe to be full of destruction because of that disintegration. Fish flew in the sky somewhere. Fire burnt hard rocks elsewhere. Beings sunk in a huge deluge. Mountains fell on top of cities and pulverized them. Wind dried up all the water. Thunder appeared everywhere. Compatibility among the root elements was completely lost. Directions disappeared as a war raged among water, fire, wind, and matter. Space seemingly disappeared.

Vaśiṣṭā turned his attention to the abode of the creator of that universe. He saw it as clearly as the Sun sees the earth at sunrise. The creator there was not shaken. He was in deep meditation. The principle Devas of that universe were with him. Enlightened beings such as Rishis were also meditating with the creator. They were in such deep meditation that they appeared to be dead.

Suddenly Vaśiṣṭā felt as if he had woken up from a dream. He remembered Brahma and the others vaguely as one remembers scenes from a dream. They had been manifestations of his own mental conditioning. Vaśiṣṭā realized that the companions of Brahma had dropped their subtle bodies. Their names and forms had disappeared from existence. They vanished in an instant from Vaśiṣṭā's sight. They had lost the materiality even of a dream-object.

Vaśiṣṭā wondered if the egosense of Brahma had vanished. Egosense creates subtle bodies and Brahma's subtle body had dissolved. Vaśiṣṭā instantaneously started seeing the dance of Rudra. Rudra is the ruler of the egosense of Brahma per Vedas.

Vaśiṣṭā saw the fearsome form of Rudra. The dissolution of universe appeared to have assumed that form. Everything was dark. Rudra's own radiance made his five faces bright. He held a trident in one hand and had nine other hands. He appeared to move in his own space because material space was lost. Rain-bearing clouds colored his body dark. His body appeared as if a huge mountain had arisen from a cosmic ocean. Rudra was himself the embodiment of that cosmic ocean.

Rishis see the disturbance of equilibrium anywhere to be a play of Rudra. Vaśiṣṭā recognized the form of Rudra because Rudra carries a trident and has three eyes. The three eyes of Rudra represent mind, intellect, and ego. Rudra is like pure space per Vedas. His color is like that of space. He is the pure and indivisible consciousness. Rishis praise him as the space-self or Akasha Ātmā. He is omnipresent because he is the space self. He is the self of all for the same reason. As the self of all he is the supreme self. His five faces represent the five sense organs of knowledge. His ten arms represent the organs of action[9] and their respective fields of action. He controls the three gunas of Prakr̥ti, the three periods, namely past, present and future. Rudra grips a trident which is a representation of three Gunas and three periods of time.

Energy moves in the infinite space of consciousness, gas stirs in placid regions in physical space, and prāṇa arises within the subtle bodies of beings because of Rudra per Vedas. Rudra is at rest only when all these movements subside. He is then Shiva the peaceful, the representation of perfect equilibrium.

Rudra began to dance then per Vaśiṣṭā's narration. Space was his stage. The process of dissolution appeared to have assumed the form of Shiva. Rudra's dance reflected the intoxication of the Cosmic waters of dissolution. Vaśiṣṭā suddenly saw a shadow behind Rudra. His mind began to wonder about the nature of that shadow. Sun did not exist at the time of dissolution to cast a shadow. The shadow stepped to the front even as Vaśiṣṭā was wondering. The shadow appeared as a female and it began to dance also. Vaśiṣṭā recognized her to be kālarātri because of the bizarre transformations in her body. She was one armed in the beginning but had several arms moments later. She threw them all to the floor as she danced. The same was the case with her feet. She sometimes danced with no feet and on several feet at other times. Her one mouth could turn into many or simply disappear. There was no predictability to the form of kālarātri.

[999] Organs of action and fields of action are both five.

Kālarātri's body swelled enormously as she danced. It encompassed heaven, earth, and the underworld. The universe reflected in her body which seemed to be like a mirror. The universe appeared and vanished randomly there. Mountains were like beads in a garland which she wore. Cities and forests were like flowers in that garland. The garland swirled as she danced vigorously. The mountains came apart from the garland and fell all over. They tore down the palaces of celestials. Mountains began to dance with kālarātri and the oceans danced on top of mountains. Continents moved around the sun which appeared sunk in water. Whole existence appeared strewn around like straw inside a mirror called kālarātri. Space was filled with different kinds of mirages. Animals appeared to be Devas. Fish held up mountains and cities.

Pairs of opposites such as day and night, creation and dissolution, purity, and impurity existed within kālarātri. Her dance had dislodged Devas who continued to look composed because of their knowledge of infinite consciousness. She danced near then far away. She shrunk infinitesimally and suddenly took on a cosmic form. Such was the manifestation of cosmic creative power.

Kālarātri's body accidentally met that of Rudra while dancing. Her body suddenly weakened. She appeared thin and transparent. Her body became robust when her dance took her away from Rudra. Her body touched Rudra's body again and she took on a cosmic form. That form instantaneously reduced to a sturdy mountain. A beautiful tree stood in its place soon. The tree also vanished at last. Kālarātri had merged into the form of Rudra.

Vaśiṣṭā began explaining the metaphor of the dance of Rudra and Kālarātri to Rama after finishing the above narration. Neither Rudra nor kālarātri danced. They were neither male nor female. They had no form. Rudra is the eternal, undivided consciousness which existed at the time of dissolution. He exists as the cause of all creation. He is the cause of all causes. He is peace or Shiva. Shiva became Rudra when the universe began to dissolve. He didn't have a form. One cannot claim that Rudra was formless either.

Vaśiṣṭā experienced the mass of consciousness which had the name Rudra. Vaśiṣṭā was the perceiver. His consciousness could perceive dissolution as a certain form. It saw dissolution in the form of a dance. Vaśiṣṭā's perception can be explained through the analogy of a word and its meaning. Neither a word nor its meaning carries significance when consciousness is absent from a perceiver. Consciousness alone recognizes the meaning and therefore gives realism to a corresponding form.

Movement is the nature of consciousness and is therefore inseparable from it. Movements within pure consciousness appeared to Vaśiṣṭā as a dance. Vaśiṣṭā considers those movements as pure because they happened in pure consciousness. There is another type of movement which originates from pure consciousness. It follows a notion arising in consciousness. It is dynamic. Vaśiṣṭā explained it to be like movements in air. Air moves as if it has a form in empty space. kālarātri is to Rudra what movement is to air. Kālarātri fulfils Rudra's will.

The dynamic energy of consciousness is prakṛti. It is undifferentiated. Consciousness is superior to this undifferentiated energy. The self of consciousness is Shiva or Rudra. Dynamic energy moves and functions as if on a cue from a momentum in the will of consciousness. It continued such movements at the time of dissolution forgetting its origin.

Kālarātri danced until reacquainting herself with Rudra. Consciousness and its dynamic energy are inseparable. Dynamic energy eventually settles back into consciousness. Rudra and Kālarātri in Vaśiṣṭā's narration correspond to the 'One' and svadhā in the nāsadīya sūktam.

4. *Word Meanings*

This chapter lists special words which are found in the nāsadīya and the other sūktams. It illustrates how these words are formed from root verbs. We can guess the intended meaning of mantras by an alternate approach when dictionary meanings fail.

We can know certain principles in a mantra better by considering a words' nigaṇṭhu/niruktam style group. This chapter illustrates this through a list of words all of which refer to space/sky. English language has one or two words to refer for example to space/sky. Sanskrit language, especially in the case of Mantras required several alternate words to refer to them. Each of them adds a special meaning to a use in a mantra. They consider a certain aspect of Sky.

व्योमन्

Vyoma is an enclosure principle. It is filled with bliss and knowledge. It has pervaded far and beyond. It can mean Sky, Heaven, Atmosphere, Space, the element Ether

Prefix वि (vi);

√अव् (av) - to drive, impel, animate

√अव् - to satisfy, refresh, promote, favor, to lead or bring to

prefix वि is an abbreviation for विशेष - in a special way

व्योमन् is विशेषेण तृप्त absolutely satisfied, or filled with bliss

Manin(?) says that it is visible to others अन्येभ्यो अपि दृष्यन्ते

व्योमन् is विशेषेण गते or व्याप्ते spread out in a unique way, nothing hides it

व्योमन् is विशेषेण ज्ञाते where all knowledge is present, filled with knowledge

√व्ये (vye) meaning to cover, clothe, wrap, envelop

√वे (ve) meaning to weave, interweave, braid, plait

to make into a cover, into a web or web-like covering, overspread as with a web

अप्रकेत

Apraketa means not discernable. It has the prefix अ which negates and the prefix प्र which means in front of. These two prefixes are added to the word केत. The word is an intransitive aspect of root चित् which has the meaning noticeable or bright.

Derivation of the word केत from the verb root चित् is shown with an example, namely, the word केतु.

केतु

Ketu means a flag. An army or a chariot is identifiable from far because of its flag. The word is an intransitive aspect of root चित् which has the meaning noticeable or bright.

चित् has the following meanings

to look at, observe, notice

intend, be intent upon

understand know

perfect tense form चिकेत meaning has understood

causative form चिकित्वान् meaning instructed

आवरीव – Something that veils or covers

contemporary form of आवरण

Prefix - आ

√वृ (vṛ) – to screen, veil, conceal

शर्मन् – something which provides shelter

√श्रि (ṣri) – be supported, fixed or dependent on

आनीत् - inhaled

अपिहित - covered

Prefix अपि

√धा

समवर्तत – originated – it was at the same level as something else

Prefix सम

√वृत् - to be, exist, become

समनुप्रविष्टः - pervaded in a delicate and uniform way

Prefix सम + anu + pra

√विश् - to enter

समेति - dissolves

Prefix सम, opposite prefix वि

√इ - go to or go towards

वयति - interleaves

√वा- to weave

ओत from same root with prefix आ

प्रोत from same root with prefix प्र

प्रसूता - birthed

Prefix प्र

√षू - to bear

अभिक्लृप्त - Having found the answer

Prefix अभि

√क्ऌप् - to put in order

नीळ - a resting place

Prefix नि

√सद् - to sit

तन्तु – offspring

Ūṇādi affix तुन्

√तन - to spread or stretch

सनि - gifts

Ūṇādi affix इनि

√षन - to give or to honor

अभिसम्बभूव - became that

Prefix अभि approach + सम् unite

√भू - to become

Words for Space and Sky

Nigantu Classifies the following as atmosphere

Ambaram – √amb (to go, to move) + (aran)

viyat – vi + √i(5) : going apart asunder

vi + √yam (kvip, tuk)

vyōma – the wide sky

barhi – bed of grass laid on yajña ground – Devas conducted the original yajña in space

dhanva – dhanv (cause to run or flow)

antarikṣam – antar + ṛkṣa (in which there are stars)

antar + īkṣa (inside of which is seen by all)

āpaḥ - √āp (pervade, obtain) + asun (ūṇādi affix)

pṛthvī - wide, broad, large, extensive, ample, abundant.

bnūḥ

svayambhūḥ - self born

adhvā – road, length

puśkaram - √puṣ (nourish), karan (ūṇādi affix)

sagaraḥ - sa + √gṛ (devour, swallow)

samudraḥ - saṃ + udra

saṃ + undi (to be wet) + raka affix

sa (short for saha) + mudrā (seal) – bounded by something

sa + muda (pleasure)

saṃ + uda (water) + √rā (to have or give) + ka affirmative

adhvaram – road leading to heaven

adhva (a road) + and ra from √rā (to give)

Nighanto classifies the following words as Mahat

Mahat – √mah (to worship) + ati (ūṇādi affix) - eminent

bradhnaḥ - the world of the Sun - √badhn + affix nak

ṛṣvaḥ - √ṛṣa (to flow, push) + sa (ūṇādi affix)

bṛhat - √bṛh (to expand, make strong)

ukṣitah - √ukṣ (get strong)

tasavaḥ

taviṣaḥ - √tu (energetic)

mahiṣaḥ - √mah (to worship) + ūṇādi affix ṭiṣac - powerful

ambhaḥ

ṛbhukṣāḥ - rab (skillful) + √kṣ (to go) + da affix

ukṣā – large - √ukṣ (to clean, to sprinkle) + ap affix

vihāyāḥ - overlooking – vi + √hā (to quit) + lyap affix

yahvaḥ - active, continuously flowing

vavakṣitha - intention

vivakṣasē – √vach (wish to speak)

√vakṣ (to grow or strong)

√vah (to hold)

ambṛṇaḥ - a vessel, roaring terribly

māhinaḥ - dominion - √mah (to worship) + inan (ūṇādi affix) + with an emphasis

gabhīraḥ - Deep – √gam (to go) + bha for ma īran (ūṇādi affix)

- Alternate form is gabhīraḥ (num affix)

kakuhaḥ - ka (wind) + √skuba (to spread), kvip affix

rabhasaḥ - fierce – √rabha (to begin) + asac (ūṇādi affix)

vrādhan – that which stirs up

virapśī – swelling, full

adbhutam – extraordinary - ata (a particle of surprise) + √bhū (to be) + uvac (ūṇādi affix)

bamhiṣṭhaḥ - most abundant – bahu (much) + iṣṭan (for superlative)

barhiṣī – ether – seat of Devas

Nighantu classifies the following as sādhāraṇa

khaḥ - hollow - √khan (to penetrate) + da affix

pṛśniḥ - √pṛś (to sprinkle) + ni (ūṇādi affix)

nākaḥ - where there is no pain – na (negation) + a (negation) + ka (happiness)

gauḥ - √gam + do (ūṇādi affix)

viṣtap – the highest part - √stambh (to fix asunder)

nabha – √nah (to bind) + bha affix

nabh (expanding, burst)

5. *Sanskrit Akṣaras*

Individual alphabets of the Sanskrit language convey existential principles. A word can convey an idea which is a combination of two or more of these principles. This chapter presents an overview of Sanskrit alphabets as codes. The chapter borrows the template of alphabet codes from an ancient text called the Lakshmi Tantra.

a	*i*	*u*	*ṛ*	*ḷ*	*ē*	*ō*	Energies
ā	*ī*	*ū*	*ṝ*	*ḹ*	*ai*	a:	

k	*kh*	*G*	*gh*	*ṅ*	Root Elements
c	*ch*	*j*	*jh*	*ñ*	Essence of elements
ṭ	*ṭh*	*ḍ*	*ḍh*	*ṇ*	Organs of action
t	*th*	*d*	*dh*	*n*	Sensory organs
p	*ph*	*b*	*bh*	*m*	Subtle Principles

Dharana/States				Brahman emanations				
y	*r*	*l*	*v*	*ś*	*ṣ*	*s*	*h*	*kṣ*

Energy emanations

anuttara	iccā	unmēṣa					bindu
a	*i*	*u*	*ṛ*	*ḷ*	*ē*	*ō*	*aṃ*
ā	*ī*	*ū*	*ṝ*	*ḹ*	*ai*	a:	*aḥ*
ānanda	īśāna	ūrjā	from iccā		Jagatyoni	Sadyojata	*visarga*

Material creation and its components

k	*kh*	*g*	*gh*	*ṅ*	Root Elements

c	*ch*	*j*	*jh*	*Ñ*	Essence in elements
ṭ	*ṭh*	*ḍ*	*ḍh*	*ṇ*	Conative organs
t	*th*	*d*	*dh*	*n*	Cognitive organs
p	*ph*	*b*	*bh*	*m*	Subtle Principles

Subtle Principles

p	*ph*	*b*	*bh*	*m*
manas/ mind	ahamkara/ ego	buddhi/ intellect	Prakriti	Puruṣa

Dhāraṇas and states of consciousness.

y	*r*	*l*	*v*
Kala	Jnana	Stamba moha	Ranjana
Kriya	Vidya	Maya	Raga
Vata	Pāvaka	Pṛthvī	Varuna

Emanations from Brahman

ś	*ṣ*	*s*	*h*	*kṣ*
Aniruddha	Pradyumna	Sankarshana	Vasudeva	
Sakti	*aiswarya*	*vijnana*		
Tejas	*virya*	*bala*		
Vyakti	*vyapti*	*udaya*	*visrama*	
(manifest)	*(pervade)*	*(arise)*	*(equilibrium)*	

Sample interpretations

parṇa – A leaf – It is a sentient (pa) entity. It converts energy (r) radiations. It creates activity or movement (ṇa) in the process – this refers to photosynthesis

> suparṇa – a bird - Its feathers reflect radiation. Feathers block sun's raditations from heating up a bird's body and makes it easy for a bird to fly.
>
> aparṇa – A name for Shakti, the divine mother - One who does not rely on energy from the outside for her activities.

arṇa – River or wave - It is in constant (a) activity or movement (ṇa). It is powered by energy (r).

prāṇa – It is associated with activity (ṇa), energises and brings joy (rā) in a sentient entity (p).

aṃbha – Source of material creation - Consciousness resonated in it in the form of nāda (aṃ). It is never fully occupied (bha). There is perfect equilibrium in it (bha). Aṃbha refers to the inexhaustible ocean which gives birth to life forms.

aṃba/aṃbu – Water. Aṃbha contains it. It expands to occupy (ba). Water within an ocean is aṃbu

bhuvana – It arises (u) from Prakṛti (bh). Consciousness goes through four states of Dharana (va) within it. It is experienced by cognitive organs (na). Bhuvana consists of material creation and sentient beings who experience it.

ātmā – It is filled with the innate bliss of Puruṣa (mā). It gains joy (ā) from the material world through sensory perception.

garbha – It mimics Prakṛti (bh) principle. Material within it is going through dynamic changes (ga). A unit of consciousness faces inwards (r) and rests in it.

ābhu – vacuum. It is filled with bliss (ā). It is in perfect equilibrium (bh). It arises from Prakṛti (bhu)

āsu – vital breath. It creates bliss (ā). It is an emanation (u) from knowingness and potential energy of Brahman (s).

śukra –It carries (u) a unit of individualized awareness (r) which is ready to be bound to the material plane of existence (k). It arises from the brilliant power of cosmic consciousness (ś).

tejas – It is refined aspect of perceptible energy (j). It interacts with sensory organs at the gross level(t)

bhrājas – It is refined part of perceptible energy (j). It interacts with the deepest source (bh) of awareness (r)

kavi – The happy one (ka) who seeks (i) turīya state of consciousness (v)

ṛca – A prayer (ṛ) connected to the essence (ca) of material existence

ṛṣi – One who prays (ṛ) or discovers prayers which seek the strength and expansionary nature of cosmic consciousness (ṣi)

6. *Hiraṇyagarbha Sūktam*

The Ambhasya pārē sūktam makes a reference to the Hiraṇya Garbha Sūktam which is the 121st Sūktam in the 10th Mandala of the Rig Veda. This chapter provides a limited analysis of this sūktam. This limited analysis is intended to add clarity to a few ideas in Chapter 1 and Chapter 2.

hiraṇyagarbhaḥ samavartatāgrē bhūtasya jātaḥ patirēka āsīt
sa dādhāra pṛthvīṃ dyāmutēmāṃ kasmai dēvāya haviṣā vidhēma

Hiraṇyagarbha was there (samavartata) having been born (jātaḥ) ahead (āgrē) of the root elements (bhūtasya). He was the One ruler (patirēka). He supported the wide expanses (pṛthvīṃ) of what we see here (imāṃ) and the subtle domain (dyām). We serve (vidhēma) the deva who is known as "Ka" with offerings (haviṣā).

The word samavartata contains the prefix "sam" which implies an equal footing to something. Brahman, or pure consciousness existed at the time of dissolution. Hiraṇyagarbha emerged as the first step out of dissolution. He was no different from Brahman. The mantra implies this idea by using the word samavartata.

Why does the mantra say that Hiraṇyagarbha was born, jātaḥ? The Nāsadīya sūktaṃ gives a clue in its third mantra. A combination of a few things emerged as a unit (jāyatēkaṃ) from undivided consciousness. Physicists call this as primordial plasma. The Nāsadīya sūktaṃ calls it a mix of primordial fluid, dark matter (tamas), and dark energy (tamas). No boundaries were discernible in it. Undivided consciousness continued to exist beyond the edge of the zone of creation. A similar idea is seen in theoretical physics. True vacuum surrounds the space of our universe. Rishis translated the image of True vacuum surrounding a sphere of space containing primordial fluids to the idea of Hiraṇyagarbha.

Garbha is a womb. Hiraṇya is golden. Hiraṇyagarbha holds together the cosmic egg. Hiraṇyagarbha refers to Brahman. The birth of Hiraṇyagarbha only refers to the appearance of a new configuration within Brahman. The mantras in this sūktaṃ refers to Hiraṇyagarbha by name "ka" from here on. The word kaḥ refers to an unknown person. kaḥ is the one who had the kāma or desire to create. The Nāsadīya sūktaṃ also says in its fourth mantra that a single desire arose at that point and became the germinal seed of the cosmic mind. kaḥ therefore refers to the unknowable creator.

Vedic texts explain why Hiraṇyagarbha got the name kaḥ with the following story. Indra once wanted to know the might of the creator. The creator told Indra that he can know the creator and his power only through an experience. The creator transferred his power to Indra for giving him an experience. Indra, after the experience, told the creator that you will be known by the name Kaḥ. Kaḥ remains a mystery without such a experience.

Brihadāraṇyaka Upanishad (1.2.1) explains the origins of Kaḥ

na ēva iha kiṃ ca na agra āsīt mṛtyu na ēva idam āvṛtaṃ ā

aśanāyayāaśanāyā hi mṛtyu tan manōakurutā ātmanvī syāmiti

sō tasyārcata āpōajāyantārcatē vai mē kam abhūditi tadēva arkasyārkatvaṃ

asmai bhavati ya ēvamētat arkasyārkatvaṃ vēda

There was nothing in the beginning. Time or death (mṛtyuḥ) that hid everything was absent. That which had not emanated (aśnāyā) was equivalent to death (mṛtyu). This unnamable (aśnāyā), Brahman had a thought in his mind "let me be (ātmanvīsyāmiti)". This thought was the worship (ārcata) from which primordial fluid (āpō) emerged. My worship or first thought has given rise to kam. This is the secret of arka. One who understands this knows the secret of Arka.

Chāndogya Upanishad 4.10.5 also explains 'kam' with the analogy of prāṇa. It also explains the letter 'kha' in relationship to the letter 'ka'. Kha is void or space or ākāśa. Ka and kha are analogous. They are infinite like Brahman. Every letter in the Sanskrit alphabet is a representation of an energy of Brahman.

> *prāṇō brahma| kaṃ brahma / khaṃ bramēti / sa hōvāca /*
> *vijānāmyaham yatprāṇō brahma, kaṃ ca tu khaṃ ca na vijānāmīti*
> *tē ha ūcuḥ yad vā va kam tadēva khaṃ yadēva khaṃ tadēva kamiti*
> *prāṇaṃ ca hāsmai tadākāśaṃ cōcūḥ*

yaḥ ātmadā baladā yasya viśva upāsatē praśiṣaṃ yasya dēvāḥ
yasya cāyāmṛtaṃ yasya mṛtyuḥ kasmai dēvāya haviṣā vidhēma

The universe obeys (upāsatē) the rules of the one who grants ātma and energy (baladā). Devas function as per the design of the one whose shadow are death and deathlessness. We serve the Deva with the name *kaḥ* with offerings.

Consciousness complements the material creation. One's ātma, as seen in the Ambhasya pāre Sūktam, is made up of pure consciousness. It is untouched by events in existence. The mantra says that the creator grants consciousness (ātmadā) to a living being. He also bestows strength (baladā) on them.

The Ambhasya pāre Sūktam calls the creator as vidhātā, the architect of the universe. The creator sets the rules. His creation follows these rules. He distributes different responsibilities to the Devas in creation. They respect it even though they have realized their oneness with the undivided consciousness or Brahman. The entire world runs per the regulations of vidhātā.

The nāsadīya sūktam tells us that death (mṛtyu) and deathlessness (amṝtam) disappeared at the time of dissolution. They reappear when Hiraṇyagarbha manifests again.

One's intellect falsely believes prāṇa which represents bala or strength to be the sustenance for one's self or ātma. Death, based on such a belief, is the separation of bala from ātma. The realization of ātma being a part of Brahman brings conviction that the ātma is the source of bala. This realization is deathlessness. Death and deathlessness are therefore related to intellect and knowledge. The above mantra therefore calls them as shadows of the creator. The next mantra introduces the idea of prāṇa.

yaḥ prāṇātō nimiśatō mahitvaika idrājā jagatō babhūva
ya īśē asya dvipadascatuṣpadaḥ kasmai dēvāya haviśā vidhēma

The creator became (babhūva) the king (rājā) of the world of beings (jagat) by powering biological activities such as breathing (prāṇātō) and blinking (nimiśatō). He rules over this (asya) domain of bipeds and quadrupeds. We serve the Deva whose name is kaḥ with offerings of haviś.

The previous mantra established the connection between the ātma and the creator. The relationship between the ātma and bala or Prāṇā is the same as the relationship between Brahman and its energy which the Nāsadīya sūktam calls as svadhā. We can therefore conclude that biological functions in a being's body are powered by the resident ātma. The next three mantras talk about the natural order among large scale entities in the universe.

yasyēmē himavantō mahitvā yasya samudraṃ rasayā sahāhuḥ
yasyēmāḥ pradiśō yasya bāhū kasmai dēvāya haviṣā vidhēma

Mountains (himavantō) are sturdy (mahitvā), the ocean (samudraṃ) is fluid (rasayā) and the directions (pradiśō) are stretched out like arms (bāhū) as per the intention of the creator. We serve the Deva who is named as kaḥ with the offerings of haviś.

yēna dyaurugrā pṛthvī ca dṛḻhā yēna svaḥ stabhitam yēna nākaḥ
yō antarikṣē rajasō vimānaḥ kasmai dēvāya haviṣā vidhēma

The atmosphere (dyau) is charged up (ugrā), the earth (pṛthvī) is solid (dṛḻhā), the sky (svaḥ) and the heaven (nākaḥ) are stable (stabhitam) as per the design of the creator. The atmosphere (antarikṣē) is filled with clouds (rajasō) per the creator's wish. We serve the Deva who is called kaḥ with haviś offerings.

yaṃ krandasī tastabhānē avasā abhyaikṣētām manasā rējamānē
tatrādhi sūra uditō vibhāti kasmai dēvāya haviṣā vidhēma

They see (abhyaikṣētām), with their inner faculty of mind (manas). They see the creator stabilizing (tastabhānē) the shining (rējamānē) and the troubled earth sky pair (krandasī) to protect (avasā) them. The Sun (sūra) rises (uditō) and illumines (vibhāti) the world under (tatrādhi) the tutelage of the creator. We serve the Deva who is named kaḥ with offerings.

The following two mantras bring a bit more clarity what the Nāsadīya sūktam tells us about creation. Hiraṇyagarbha is the witness (adhyakṣa). He witnessed the birth of the Devas, Yagna, and prajāpati. We can conclude per these two mantras that the entity who arose at the beginning of creation and whom the Nāsadīya sūktam does not name is indeed Hiraṇyagarbha.

āpō ha yadbṛhatīr viśvamāyan garbhaṃ dadhānā janayantīr agniṃ
tatō dēvānām samavartatāsurēkaḥ kasmai dēvāya haviṣā vidhēma

Primordial fluids (āpō) had spread (viśvamāyan) over wide expanses (bṛhatīr) of the universe. They did so to give birth (janayantīr) to Agni from the womb which they carried (dadhānā). There then existed (samavartata) the one sap (āsu) of all the devas. We serve the Deva who is called kaḥ with the offering of haviś.

The Nāsadīya sūktam, in its fifth mantra, tells us that the primordial fluids broke up into three parts. The three parts were (a) radiation, (b) germinal seeds of beings and (c) the root elements of material creation. The mantra above talks about one part of that idea. Devas who are also beings were born at that point. The above mantra mentions the name of only one Deva, namely Agni. This however is a reference to all Devas.

Hiraṇyagarbha took on an additional role (samavartata) as the specialty in all Devas. Devas have special powers because of this influence. The first mantra of this sūktam suggests the idea that Hiraṇyagarbha is a name given to Brahman. The name Hiraṇyagarbha corresponds to the manifestation of a new configuration within Brahman and to the addition of a new role. The first mantra of this sūktam affirms that it is indeed a role addition by using the word "samavartata".

The present mantra suggests another role addition by using the word "samavartata". Hiraṇyagarbha gets the name prajāpati per this new role. The name Prajāpati makes sense only after the birth of beings such as Devas. Vedic mantras therefore see Prajāpati as the being within the cosmic egg, within the primordial fluids.

yascidāpō mahinā paryapaśyaddakṣaṃ dadhānā janayantiryajñam
yō dēvēṣvadhi dēva ēka āsīt kasmai dēvāya haviṣā vidhēma

The primordial fluids (āpō) carried (dadhānā) dakṣaṃ through their might (mahinā) and gave birth (janayantir) to yajña. Hiraṇyagarbha witnessed (paryapaśyad) this. The Deva who is named kaḥ is above all Devas. We serve him with the offerings of haviś.

Primordial fluids (āpō) originated at the time of dissolution. They appeared directly from pure consciousness. They are of the nature of the power of Brahman. The nāsadīya sūktam tells us that Brahman alone existed at the time of dissolution along with the undivided energy named svadhā. Creation began as a perturbance in svadhā the energy field. The Nāsadīya sūktam tells us that the first transformation in that energy field appeared as the primordial fluids (salila). Salila is the primordial plasma per theoretical physicists

The nāsadīya sūktam tells us that the germinal seeds of beings and the root elements of material creation arose from svadhā. These were only the building blocks. A seed must be nurtured into a tree to perceive the potentiality within it. Complex structures must be designed from simple elements to create a beautiful and functional dwelling. Skill is needed for that. The word Dakśa refers to skill. Primordial fluids carried skills in the womb of creation. Dakśa is Prajāpati. Vedic mantras picture Prajāpati to be within the cosmic egg, in the primordial fluids.

Yajña is the process which created diversity in creation. The primordial fluids birthed Yajña. Devas are the performers of cosmic Yajñas. Devas and Rishis help Prajāpati in Yajñas. Human beings conduct rituals which are also called as Yajña. These rituals mimic the cosmic Yajña of yore. Yajñas of humans are named so because of it. Vedic literature describes this through stories. These stories expand the base idea in the above mantra.

The above mantra reminds us that there is one and only Deva despite Vedic rituals adoring a diversity of Devas. Vedic priests chant the mantras in hiraṇyagarbha sūktam while making offerings to the One Deva. Each mantra in the hiraṇyagarbha sūktam ends with a phrase which affirms the reality of a single unknowable Deva. The following mantra of hiraṇyagarbha sūktam is a prayer to kaḥ, the highest Deva for protection from harm. The mantra tells us that this Deva is himself the creator of the primordial fluids!

mā nō hiṃsījjanitā yaḥ pṛthivyā yō vā divaṃ satyadharmā jajāna yascāpaścandrā bṛhatīrjajāna kasmai dēvāya haviṣā vidhēma

7. Uttara-nārāyaṇa Sūktam

The ambhasya pārē sūktam refers to this sūktam which is a part of the Taitrīya āraṇyaka (prapāṭaka-3 anuvākā-13). This chapter provides a limited overview of the Uttara Nārāyaṇa sūktam to provide clarity to a few ideas in Chapter 1 and Chapter 2.

adbhya saṃbhūtaḥ pṛthivyai rasāsca viśvakarmaṇaḥ samavartatādi tasya tvaṣṭā vidadhāt rūpam ēti tat purūśasya viśvamājanaṃ agrē

Puruṣa arose from the primordial waters and subsequently again from refined materials (pṛthivyai rasāsca). He arose under the supervision (ādi) of viśvakarma. His form (rūpam) is this universe (viśvam). Tvaṣṭā, arrived (ēti) in the universe which was manifesting all over (ājanaṃ) to give its shape (vidadhāt rūpam).

This mantra uses the word "samavartata" as the mantras in hiraṇyagarbha sūktam do. Puruṣa is only a name and not a new emanation. It is used to reference the cosmic consciousness within the advancing creation. The mantra therefore says that Puruṣa arose twice. He arose when the universe was nascent. Purusha arose also when the universe was filled with diverse materials. This mantra recognizes the evolutionary nature of the universe. Puruṣa is a continuation.

The above mantra provides us a time reference after which the name Puruṣa applies for cosmic consciousness. It says that Puruṣa arose under the supervision of Viśvakarma, or vidhātā, the cosmic architect. Desire, activity, and awareness are three dimensions of consciousness. The nāsadīya sūktam tells us that a single desire arose in creation in the very beginning. This was the germinal seed for the cosmic mind.

Viśvakarma is the cosmic architect. He is the counterpart to the cosmic mind which sets activities in motion in the universe.

The following mantra from Rigveda defines the role of Viśvakarma. It says that Viśvakarma is the non-mind who is the vigor behind activities in the universe. It says that the play of the primordial waters and the seven rishis originate in him.

> viśvakarmā vimanā ādvihāyā dhātā vidhātā pramōta saṃdṛk
> tēṣamiṣṭāṇi samiṣā madanti yatrā saptaṛṣīnpara ējamāhuḥ

Tvaṣṭā is the Deva of forms. Tvaṣṭā appeared when the material universe which is the body of Puruṣa, was ready to manifest. The following mantra defines the role of Tvaṣṭā. Tvaṣṭā is called viśvarūpaḥ, the cosmic form. He is the vital force of Devas

> *dēvastvaṣṭa savitā viśvarūpaḥ pupōśa prajāḥ purudhā jajāna*
> *imā ca viśvā bhuvanānyasya mahaddēvānāmasuratvamēkam*

Scientists attribute the largescale structures in the present universe to regions of dark matter in the early universe. Quantum physicists say that quantum fluctuations created perturbations within a tiny bubble of vacuum. These perturbations magnified when the bubble inflated to create the space of the universe. Dark matter regions in the present universe can be mapped back to the initial quantum fluctuations. Rishis too thought about the reason for the structure of the material creation. The nāsadīya sūktam suggests that dark matter existed in the universe which was filled with primordial fluid. Stories about Tvaṣṭā, his son, and Devas provide a few related hints. These hints suggest that a delicate equilibrium between opposing forces had sustained the creation process.

The next sūktam is about Puruṣa and consciousness. The seer who contemplated on the nature of Puruṣa declared the following

vēdāhamētaṃ puruṣaṃ mahāntaṃ āditya varṇaṃ tamasaḥ parastāt
tamēvam vidvān amṛta iva bhavati nānyaḥ pamthā ayanāya vidyatē

I know the mighty (mahāntaṃ) Puruṣa who is bright like the sun (āditya varṇaṃ). He is beyond the reach of anyone who is associated with ignorance (tamas). One attains deathlessness by knowing the nature of Puruṣa. There is no other path to reach deathlessness.

The idea of Puruṣa is difficult to understand. The above mantra attributes this difficulty to ignorance about the real nature of existence. It seems reasonable to consider consciousness to be a by-product of chemical reactions in the brain and the body. It is more difficult to accept the idea that the human body is a temporary residence for consciousness which is indestructible.

Cause and effect is the rule in material creation. Body becomes weak without food. One becomes unconscious when the weakness continues for a while. Food can be therefore mistaken to be the source of consciousness. It is however ignorance to refuse to look beyond such cause-effect observations.

Science studies phenomena from the angle of cause and effect. Science however is unable to explain every phenomenon in nature using a simple cause and effect model. Science uses chance-based explanations in such cases.

Is creation purely a chance or is it based on intelligence, per science? Vigorous debates are raging now in new fields of science such as cosmology and theoretical physics. Rishis answered this age-old question by placing consciousness at the root of creation. They left a place for chance in their model too.

One's ātmā is the seat of consciousness within the human body. Three states of consciousness, namely, waking, sleep and dream arise from the ātmā. Body decays when the ātmā disassociates from it. All ātmās rest in Puruṣa. Puruṣa continues to exist when the universe dissolves. Puruṣa is cosmic consciousness. Hiraṇyagarbha is another name for Puruṣa. Prajāpati is his name also.

Consciousness does not follow set concepts which can explain the rest of existence. The following mantra which talks about the nature of consciousness presents two seemingly conflicting ideas.

prajāpatiḥ carati garbhē antaḥ ajāyamānō bahudā vijāyatē
tasya dhīrāh parijānanti yōniṃ marīcīnām padamiccanti vēdasaḥ

Prajāpati moves (carati) within the womb (garbhē antaḥ). He is not born (ajāyamāna). He is born (vijāyatē) in a variety of ways (bahudā). The determined (dhīrāh) ones understand (parijānanti) his origin (yōniṃ). The knowers (vēdasaḥ) seek (iccanti) the status (padamiccanti) of the Marīcī sages.

Hiraṇyagarbha holds the womb of creation. The unborn Prajāpati moves inside the womb. The universe arises from the womb. The universe is the body of Puruṣa. Can we therefore not say that Puruṣa took birth when the universe came into view? Puruṣa and Prajāpati, the creator are merely names for cosmic consciousness which has existed from before. Ideas in this mantra thus appear incongruent. The truth about Puruṣa lies in them. Truly, only dhīrāh, or the bold ones can find congruency among them. It is a difficult task. The vēdasaḥ, the knowers, have succeeded in sorting through these ideas with the guidance of a Guru and through personal experiences.

Intention is the seed of every conscious transformation. Attention fuels an action to fruition. The degree of success in any action depends on the clarity of intention and on the quality of attention. Intention and attention both relate to consciousness. One needs the clarity of Prajāpati to mimic the success of Prajāpati in creating the universe. One with knowledge about Puruṣa gains the clarity of Prajāpati. Vedic texts list the names of prominent Rishis who were the equals of Prajāpati and who helped Prajāpati. Foremost among such Rishis are the Marīcis.

One who gains even an insight into the nature of Puruṣa gains wisdom to influence the surroundings. The next mantra summarizes the nature of Puruṣa with a few phrases

yō dēvēbhyō ātapati yō dēvānām purōhitaḥ
pūrvō yō dēvēbhyō jātaḥ namō ṛcāya brāhmayē

He stimulates (ātapati) the Devas. He is the benefactor (purōhitaḥ) of the Devas. He was born (jātaḥ) before (pūrva) the Devas. Salutations to Him who is in the form mantras (ṛcāya) which contain the knowledge of Brahman (brāhmayē).

The knower, the known and knowledge merge in Brahman. The knower becomes ecstatic when such a merger happens. Vedic mantras talk about only one reality which has taken different forms. Veda itself is of the nature of Brahman.

The Rishi expresses his ecstasy with the word namō in the above mantra. The word indicates that the seer's mind has merged into the all-pervading One while contemplating on the nature of Puruṣa. The next mantra hints at the secret origins of Vedic mantras

ṛcam brāhmaṃ janayantaḥ dēva agrē tadabruvan
yastvaivam brāhmaṇō vidyāt tasya dēvā asan vaśē

Devas played a part in (janayantaḥ) the generation of the knowledge of Brahman (brāhmaṃ). This knowledge is in the form of the Vedas (ṛcam). Devas generated it in the beginning of creation (agrē). Devas oblige the knower (brāhmaṇa) of Brahman who knows this secret.

Rishis cognized the Vedic mantras in their meditations millennia ago. They recorded the names of Devas who helped cognize the corresponding mantra. Devas were participants in the process of discovering mantras from the field of undivided consciousness. Mantras contain clues about the mysteries in universe. One of the biggest mystery in creation is the connection between the subtle and the gross planes of existence. This connection exists because these two planes share a common origin, namely the field of undivided consciousness, the Brahman.

The origin of speech as a mode of human communication is a mystery. Mantras are a medium of human communication at its subtlest refinement. One gains a mastery over mantras through special training which brings out certain fine skills. The title "Brāhmaṇa" was historically given to anyone who committed to developing such skills. A brāhmaṇa helps in preserving the mantras for the next generation.

Vedas survived because of the effort of brāhmaṇas in the past centuries. A brāhmaṇa, however, may not understand the mystery behind every mantra. The one who understands the nature of the phrase "ṛcam brāhmaṃ" which refers to the knowledge of Brahman are special. Devas become the servants of such a special person.

hṝsca tē lakṣmīśca patnyau ahō rātrē pārśvē
nakṣatrāṇi rūpam aśvinau vyāttam
iśṭaṃ maniṣāna amuṃ maniṣāna sarvaṃ maniṣāna

Your consorts are Hṝ and Lakṣmī. Day and night are by your side. Objects in creation which illuminate (nakṣatrāṇi) are parts of your body. Devas are important organs in your body. The aśvini pair is your open mouth. Grant us knowledge which is conducive (iśṭaṃ), bestow material prosperity (amuṃ), and bless us with all comforts.

We learned from the nāsadīya sūktam that svadhā is the name of pristine energy. That energy was an integral part of Brahman at the beginning of creation. The energy of Brahman differentiated as creation progressed. Rishis recognized this and coined different names for distinct energies. The above mantra names two energies which are associated with Puruṣa. These two energies are Hṝ and Lakṣmī. They are a part of existence. Vāṇī, per Vedic tradition, is the name of the creative energy which is associated with hiraṇyagarbha.

8. Sāyaṇā's Commentary

Sāyaṇācārya's commentary on the Vedas has served as a reference for Vedic scholars for several centuries. Sāyaṇācārya has presented mantras in a context of their application in Yagnas. Sāyaṇācārya was associated with the Vijayanagar kingdom in South India. His commentaries were popular all over India when the British began their rule in India. Colonial scholars used Sāyaṇācārya's commentaries to translate the Vedas into western languages. Sāyaṇācārya had gathered knowledge from various manuscripts which were available to him in the fourteenth century.

This chapter presents the text of Sāyaṇācārya's commentary on the Nāsadīya Sūktaṁ in Sanskrit language. It illustrates how fourteenth century understanding of mantras can be adapted to a modern context. The chapter presents short explanations in English for each segment of the original commentary in Sanskrit. Short explanations illustrate the style and tenor of interpretations which were popular during medieval times. They however suffice as a potential framework for understanding mantras in a context of modern science. The following are a few examples

- Sāyaṇācārya interprets the phrase "vyoma para" as fourteen domains of existence. Cosmology per Puranas mentions fourteen domains of existence. Cosmology per Yoga Vaśiṣṭā however mentions three kinds of space. The theme in Yoga Vaśiṣṭā better reflect the semblance of ideas in the nāsadīya Sūktaṁ to Big Bang theory.

- Sāyaṇācārya interprets the 'One" who existed at the time of dissolution to be a reference to the Cosmic Being. The word "One" also refers to undivided consciousness. The idea of the energy of vacuum in theoretical physics and the idea of "pure" consciousness per Yoga Vaśiṣṭā begin to unite in the context of the word "One" meaning undivided consciousness.
- Sāyaṇācārya interprets the word "rashmi" to be a reference to activities related to creation. Those activities spread out quickly, like the brightness at dawn. The word rashmi also refers to radiation. The associated mantra can be seen as a reference to CMB phenomenon in the context of the word Rashmi referring to radiation. Such an interpretation also highlights the parallel between the primordial fluid in the cosmic egg (Brahmāndā) and the primordial plasma of a scientist.
- Sāyaṇācārya takes the word "rētōdhā" or germinal seeds of beings to be impression related to past karmas, namely, from an earlier kalpa cycle. It is equally plausible to take the word "rētōdhā" or the germinal seeds of beings to be the beginning of individual minds. Yoga Vaśiṣṭā's explanations about the universal nature of consciousness and the limited nature of individual mind fit perfectly well with the latter approach.

A full translation of Sāyaṇācārya's commentaries maybe valuable only to someone interested in Hindu philosophy from that time. A serious researcher however may want to refer to the original commentary in full to understand other parallels between the mantras presented in this book and modern science. I decided to include the original commentary of Sāyaṇācārya at least in Devanagari script for the serious researchers. Explanations in the first chapter of this book fit SriSri's interpretation of mantras. I have included SriSri's interpretations in this Chapter to show the scientific orientation in his explanations which continues to inspire my research on ancient literature.

In this Chapter, I have shown each mantra in its transliterated format, followed by its split-word format in Devanagari and further followed by SriSri's interpretation. I have included Sāyaṇā's commentary in Sanskrit along with a broad translation of the commentary after SriSri's interpretation for each mantra.

ṇasadasinnō sadāsīttadānīm nasīdrajō nō vyōmā parō yat
Kimāvarivaḥ kuha kasya śarmannambhaḥ kiṃāsīd gahanaṃ gabhīram

न । असत् । आसीत् । नो इति । सत् ।आसित् । तदानीम् ।

न । आसीत् । रजः । नो इति । विऽओम । परः । यत् ।

किम् । आ । अवरीवरिति । कुह । कस्य । शर्मन् । अम्भ । किम् । आसीत् । गहनम् । गभीरम् ।

At first was neither Being nor Non-being. There was not air nor yet sky beyond. What was wrapping it? Where? In what protection? Was water there, unfathomable and deep?

'तपसस्तन्महिनाजायैकम्' इत्यादिनाग्रे सृष्टिः पर्तिपादिष्यते । अधुना ततः प्रागवस्था निरस्तसमस्तप्रपञ्चा या प्रलयावस्था सा निरूप्यते ।

The above mantra describes the state of the universe before its birth. This conclusion is based on the phrase within quotes occurring in a later mantra of the Sūktaṁ. That phrase refers to the birth of the universe.

*तदानीं प्रलयदशायामवस्थितं यदस्य जगतो मूलकारणं तत् *असत् शशविषाणवन्निरुपाख्यं न आसीत् । न हि तादृशात् कारणादस्य सतो जगत उत्पत्तिः संभवति। तथा *नो *सत् नैव सदात्मवत् सत्वेन निर्वाच्यम् *आसीत्। यद्यपि सदसदात्मकम् प्रत्येकम् विलक्षणं भवति तथापि भावाभावयोः सहावस्थानमपि संभवति। कुतस्तयोः तादात्म्यमिति उभय विलक्षणमनिर्वाच्यमेवासीदित्यर्थः।

Existence (bhāva) and Nullity (abhāva) were absent as much as sat and asat were absent.

ननु नो सदिति पारमार्थिकसत्वस्य निषेधः। तर्हात्मनोऽप्यनिर्वाच्यत्वप्रसङ्गः। अथोच्येत। न । आनीदवातमिति तस्य सत्वमग्रे वक्ष्यते परिशेषान्मायाया एवात्र सत्वं निषिध्यत इति। एवमपि तदानीमिति विशेषणानर्थक्यं व्यवहारदशायामपि तस्याः पारमार्थिकसत्वाभावात्। अथ व्यावहरिकसतां पृथिव्यादीनां भावानां विद्यमानत्वात् कथं नो सदिति निषेधः। तत्राह । नासीद्रज इत्यादि। 'लोका रजांस्युच्यन्ते' *(निरु. ४.१९)* इति यास्कः। अत्र च सामान्यापेक्षमेकवचनम् ।

Sattva Guna which is associated with Māyā was absent. We cannot however conclude that Sattva associated with the supreme self (paramātmā) was absent.

व्योम्नो वक्ष्यमाणत्वात्तस्याधस्तनाःपातालादयःपृथिव्यन्ता नासन्नित्यर्थः। तथा व्योम अन्तरिक्षं तदपि नो नैवासीत् । पर इति सकारान्तं परस्तादित्यर्थे वर्तते। परशबाच्छान्दसोस्तातेरर्थेसिप्रत्ययः। पर व्योम्नः परस्तादुपरिदेशेद्युलोकप्रभृतिसत्वलोकान्तं यत् अस्ति तदपि नासीदित्यर्थः।

अनेन चतुर्दशभुवनगर्भं ब्रह्माण्डं स्वरूपेना निषिद्धं भवति। अथ तदावकत्वेन पुराणेषु प्रसिध्दानि यानि वियदादिभूतानि तेषामवस्थानप्रदेशं तदावरणनिमित्तं चाक्षेप मुखेनक्रमेना निषेधयति किमावरीविरिति।

Fourteen domains of existence were missing. Five root elements occupy these fourteen domains of existence.

किम् आवरणीयं तत्वमावरकभूतजातम् आवरीवः। अत्यन्तमावृणुयात्। आवार्याभावात् तदावरकमपि नासीदित्यर्थः॥ वृणोतेर्यङ्लुगन्ताच्छान्दसे लङि तिपि रूपमेतत्॥ यद्वा। किमिति प्रथमैव। किं तत्वमावरकमावृणुयात् । आव्रियमाणवत्तदपि स्वरूपेण नासीदित्यर्थः । आवृण्वत् तत्तत्वं *कुह कुत्र देशेवस्थायावृणोति । आधारभूतस्ताद्दशो देशोपि नासीदित्यर्थः॥ किंश्ब्दात् सप्तम्यर्थे हप्रत्ययः। 'कु तिहोः ' (पा. सू. ७.२.१०४) इति प्रकृतेः क्वादेशः॥ *कस्य *शर्मन् कस्य वा भोक्तुर्जीवस्य शर्मणि सुखदुःख साक्षाभावानामुपभोगार्थं हि सृष्टिः। तस्याम् हि सत्यां ब्रह्माण्डस्य भूतैरावरणं प्रलयदशायां च भोक्तारो जीवा उपाधिविलयात् प्रलीना इति कस्य कश्चिदपि भोक्ता न संभवतीत्यावरणस्य निमित्ताभावादपि तन्न घटत इत्यर्थः। एतेन भोग्यप्रपञ्चोऽपि तदानीं नासीदित्युक्तं भवति ॥ किंशब्दादुत्तरस्य ङ्सः 'सावेकाचः' इति प्राप्तस्योदात्तत्वस्य 'न गोश्वन्साववर्णः' इति प्रतिषेधः। 'सुपां सुलुक्' इति शर्मणः सप्तम्या लुक् ।

Creation has two parts, namely, a part that enjoys the creation and a part that is enjoyed. Beings constitute the former. These beings were absent at the time of dissolution. A wrapper could not have hidden the universe then. An idea of a wrapper presupposes the idea of someone upholding it.

यद्यपि सावरणस्य ब्रह्माण्डस्य निषेधेन तदन्तर्गतमप्सत्त्वमपि निराकृतं तथापि आपो वा इदमग्रे सलिलमासीत्' (तै. सं. ७.१.५.१) इति श्रुत्या कश्चित्दपां सद्भावमाशङ्केत । तं प्रत्याचष्टे अम्भः किमासीत् इति । *गहनम् दुष्प्रवेशं *गभीरं दुरवस्थानमत्यगाधम् ईद्दशम् *अम्भः *किमासीत् । तदपि नैवासीदित्यर्थः। श्रुतिस्त्ववान्तरप्रलयविषया ॥

The source of Sattva, namely primordial Prakr̥ti (ambha) was absent. Sattva which fills cosmic egg was therefore missing then.

ṇa mṛtyurāsīt amṛtam tarhi na rātriyā anha asīt prakētaṃ ānīt avātaṃ svadayā tadēkaṃ tasmāt ha anyanna para kiñcanāsa

न । मृत्युः । आसीत् । अमृतम् । न । तर्हि । न । रात्र्याः । अन्हः । आसीत् । प्रSकेतः ।

आनीत् । अवातम् । स्वधया । तत् । एकम् ।

तस्मात् । ह । अन्यत् । न । परः । किम् । । चन आस ।।

There was no death then, nor yet deathlessness; of night or day, there was not any sign.

The One breathed without breath, by its own impulse. Other than that was nothing else at all.

ननूक्तस्य प्रतिसंहारस्य संहर्त्रपेक्षत्वात् स एव संहर्ता मृत्युर्विद्यत इत्यत आह *न *मृत्युरासीत् इति । ननु यदि स नासीत् तर्हि तदभावकृतम् *अमृतम् अमरणं प्राणिनामवस्थानां तदानीमपिस्यात् तत्राह । *अमृतं *न *तर्हि इति। तर्हि तस्मिन् प्रतिहारसमये । अयं भावः। सर्वेषां प्राणिनां परिपक्वं भोगहेतुभूतं सर्वं कर्म यदोपभुक्तमासीत् तदा भोगाभावान्निष्प्रयोजनमिदं जगदिति

परमेश्वरस्य मनसि संजिहीर्षा जायते । तथैव स मृत्युः सर्वं जगत् संहरत इति किमनेन मृत्युना संहर्त्रा तदभावकृतं वा कथममरणं स्यादिति।

Creatures who existed in the earlier universe had exhausted their respective Karmas. The supreme self therefore decided to dissolve that universe with the help of Death (mr̥tyu).

एतदेवाभिप्रेत्य कठैराम्नायते - यस्य ब्रह्म च क्षत्रं चोभे भवत ओदनः॥ मृत्युर्यस्योपसेवनं क इत्था वेद यत्र सः' *(क.उ २.२४)* इति। नन्वेतस्य सर्वस्याधिकरणभूत्ः कालो विद्यत इत्यत आह न रात्र्या इति । *रात्र्याः *अह्नः च *प्रकेतः प्रज्ञानं *न *आसीत्। तद्धेतु भूतयोः सूर्याचन्द्रमसोरभावात् । एतेनाहोरात्रनिषेधेन तदात्मको मासर्तुसंवत्सरप्रभृतिकः सर्वः कालः प्रत्याख्यातः। कथं तर्हि नो सदासीत्तदानीमिति कालवाची प्रत्ययः। उपचारादिति ब्रूमः। यथेदानींतननिषेधस्य कालोऽवच्च्छेदकस्तथा मायापि तदव्च्छेदहेतुरित्यवच्छेदकत्वसाम्येनाऽकालेपि कालवाची प्रत्ययः।

Time principle was absent. The phrase 'day was covered by night' refers to the time principle.

यदवादिष्म ब्रह्मणः परमार्थसत्वमग्रे वक्ष्यत इति तदिदानीं दर्शयत्यानीदिति। *तत् सकलवेदान्तप्रसिध्दं ब्रह्मतत्वम् *आनीत् प्राणितवत् । नन्वेवं प्राणनकर्तुजीवाभावापन्नस्यैव ब्रह्मणः सत्वं स्यात् न विवक्षितस्य निरुपाधिकस्य ब्रह्मणः। अप्राणो ह्यमनाः शुध्दः' इति तस्य प्राण स्ंबन्धाऽभावात् तत्राह आनीदवातमिति। अयमाशयः।

Sattva associated with the supreme self was present. The word 'inhaled (ānīt)' attests to this. The supreme self is independent of prāṇa. The phrase 'without air (avātaṁ)' shows this independence.

आनीदित्यत्र धात्वर्थक्रिया तत्कर्ता तस्य च भूतकालसंबन्ध इति त्रयोर्था प्रतीयन्ते। तत्र समुदायो न विधीयते यथाग्नेयोऽष्टाकपाल इति येन ब्रह्मणः सत्वं न स्यात् । किं तर्ह्यनेन कर्तृत्वमनूद्य भूतकालसत्तालक्षणो गुणो विधीयते दध्ना जुहोतीति वाक्यान्तरविहिताग्निहोत्रानुवादेन तत्र गुणविधानम्। तत्राप्यनेन कर्तृत्वविशिष्टस्य न पूर्वकालसत्ता विधीयते तन्निषेधानुपपत्तिप्रसङ्गात् अतोनेन कर्तृत्वेन इदानींतनोपलक्षितं यन्निरुपाधिकं परं ब्रह्म तस्यैव भूतकालसत्ता विधीयत इति न कश्चिदोष इति।

We cannot take the word ānīt as implying the supreme self to be the actor.

नन्वीदृशस्य ब्रह्मणो मायया सह संबन्धासंभवात् सांख्याभिमता स्वतन्त्रा सद्रूपा सत्वरजस्तमोगुणात्मिका मूलप्रकृतिरेवाभिमतेति कथं नो सदिति निषेधः। तत्राह *स्वधया इति । स्वस्मिन् धीयते ध्रियत आश्रित्य वर्तत इति स्वधा माया । तया तद्ब्रह्मैकमविभागापन्नमासीत्। 'सहयुक्तेऽप्रधाने' (पा.सू २.३.१९) इति तृतीया सहशब्दयोगाभावेऽपि सहार्थयोगे भवति 'वृध्दो यूना' (पा.सू १.२.६५) इति निपातनाल्लिङ्गात् । अत्र प्रकृतिप्रत्ययाभ्यां तस्याः स्वातन्त्र्यं निवार्यते।

Svadhā or māyā was the cause of inhalation. Svadhā is not separate from Brahman. Svadhā however can act independent of the supreme self when creation has begun.

यद्यपि असङ्गस्य ब्रह्मणस्तया सह संबन्धो न संभवति तथापि तस्मिन्नविद्यया तत्स्वरूपमिव संबन्धोऽप्यध्यस्यते यथा शुक्तिकायां रजतस्य । एतेन सद्रूपत्वमपि तस्याः प्रत्याख्यातम् । ननु यदि माया ब्रह्मणा सहाऽविभागापन्ना तर्हि तस्या अनिर्वाच्यत्वात् ब्रह्मणोऽपि तत्प्रसङ्ग इति

कथं तस्य सत्वमुक्तम् आनीदवातमिति। ब्रह्मणो वा सत्वासत्तस्या अपि सत्वप्रसङ्ग इति कथं नो सदासीदिति सत्वप्रतिषेधः। मैवम्। अयुक्तिदृष्ट्यैक्यावभासेऽपि युक्त्या विवच्य मायांशस्यानिर्वाच्यत्वं ब्रह्मणः सत्वं प्रतिपादितम्।

Mother of pearl appears like silver. The relationship of Avidya and Brahman is a similar illusion.

ननु दृग्दृश्याविति द्वावेव पदार्थौ आनीदवातं स्वधयेति तौ चेङ्गीक्रियते तत्किमपरमवशिष्यते यत् नासीद्रजः इत्यादिना प्रतिषिध्येत तत्राह तस्मादिति। *तस्माध्द तस्मात् खलु पूर्वोक्तान्मायासहितात् ब्रह्मणः *अन्यत् *किं *चन किमपि वस्तु भूतभौतिकात्मकं जगत् *न *आस न बभूव ॥ छन्दस्युभयथा' इति लिटः सार्वधातुकत्वादस्तेर्भूभावाभावः। ननु तदानीमन्यस्य सत्वनिषेधो न शङ्क्यः। असत्वे चाप्रसक्तत्वान्न' निषेधोपयोग इत्यत आह पर इति। *पर परस्तात् सृष्टेरूर्ध्वं वर्तमानमिदं जगत् तदानीं न बभूवेत्यर्थः। अन्यथा उक्तरीत्या क्वचिदपि निषेधो न स्यादिति भावः॥

The universe which we see now was absent then. The phrase 'rajas was absent at that time' attests to this.

tama āsīt ṭamasā gul̲hamagrē aprakētaṃ salilamā sarvagm idaṃ tuccyēna abhvapihitaṃ yadāsīt tapasas tan mahinā jāyataikaṃ

तमः । आसीत् । तमसा । गूळ्हम् । अग्रे । अप्रऽकेतम् । सलिलम् । सर्वम् । आः इदम्

तुच्छ्येन । आभु । अपिऽहितम् । यत् । आसीत् ।

तपसः । तत् । महिना । अजायत । एकम् ।।

Darkness was there all wrapped around by darkness, and all was water indiscriminate. All was energy indiscriminate.

Then that which was hidden by the void, the one emerging, stirring, through power of Ardor, came to be.

ननूक्तप्रकारेण यदि पूर्वमिदं जगन्नासीत् कथं तर्हि तस्य जन्म । जायमानस्य जनिक्रियायां कर्तृत्वेन कारकत्वात् कारकं च कारणावान्तरविशेष इति कारकस्य सतो नियतपूर्वक्षणवर्तित्वस्य अवश्यं भावात्। अथैतहोपपरिजिहीर्षया जनिक्रियायाः प्रागपि तद्विद्यत इत्युच्यते।

How was the universe born? There was no cause for it.

कथं तस्य जन्म। अत आह तमसा गूळ्हमग्रे इति । *अग्रे सृष्टेः प्राक् प्रलयदशायां भूतभौतिकं सर्वं जगत् *तमसा *गूळ्हम् । यथा नैशं तमः सर्वपदार्थजातमावृणोति तद्वत्। आत्मतत्वस्यावरकत्वान्मायापरसंज्ञं भावरूपाज्ञानमत्र तम इत्युच्यते। तेन तमसा निगूढं संवृतं कारणभूतेन तेनाच्छादितम् भवति। आच्छादकात् तस्मात्तमसो नामरूपाभ्यां यदाविर्भवनं । तदेव तस्य जन्मेत्युच्यते । एतेन कारणावस्थायामसदेव कार्यमुत्पद्यते इत्यसद्वादिनोऽसत्कार्यवादिनो ये मन्यन्ते ते प्रत्याख्याताः।

The universe of forms and names arose out of tamas or ignorance. Those who consider asat to be the cause of the universe use the last statement as a support for their argument.

ननु कारणे तमसि तज्जगदात्मकं कार्यं विद्यते चेत् कथं नासीद्रज इत्यादिनिषेधः। तत्राह *तम *आसीत् इति । तमो भावरूपाज्ञानं मूलकारणम् । तद्रूपता तदात्मनाम् । यतः सर्वं जगत् प्राक् तम आसीदतो निषिध्यत इत्यर्थः। नन्वावरकत्वादावरकं तमः कर्तृ आवार्यत्वाज्जगत्कर्म । कथं तयोः कर्मकर्त्रोस्तादात्म्यम्। तत्राह अप्रकेतमिति । * अप्रकेतम् अप्रज्ञायमानम्।

अयमर्थः। यद्यपि जगतस्तमसश्च कर्मकर्तृभावोयौक्तिको विद्यते तथापि व्यवहारदशायामिव तस्यां दशायां नामरूपाभ्यां विस्पष्टं न ज्ञायत इति तादात्म्यवर्णनम् ॥

Forms and names were not recognized because of ignorance.

अत एव मनुना स्मर्यते आसीदिदं तमोभूतमप्रज्ञातमलक्षणम्। अप्रतर्क्यमनिर्देश्यं प्रसुप्तमिव सर्वतः' *(मनु १.५)* इति । कुतो वा न प्रज्ञायते तत्राह। *सलिलम्। 'षल गतौ'। औणादिक इलच्। * इदम् दृश्यमानं * सर्वं जगत् सलिलं कारणेन संगतमविभागापन्नम् *आः आसीत् । अस्तेर्लङि तिपि 'बहुलं छन्दसि' इतीडभावे 'हल्ङ्याब्भ्यः' इति तिलोपे 'तिप्यनस्तेः' *(पा.सू. ८.२.७३)* इति पर्युदासाद्दकाराभावः। यद्वा सलिलमिति लुप्तोपमम् । सलिलमिव । यथा क्षीरेणाविभागापन्नं नीरं दुर्विज्ञानं तथा तमसाविभागापन्नं जगन्न शक्यविज्ञानमित्यर्थः।

Fixed boundaries are not seen in a fluid. The universe being immersed in tamas was similarly unrecognizable. This is like water mixed with milk.

ननु विविधविचित्ररूपभूयसः प्रपञ्चस्य कथमतितुच्चेन तमसा क्षीरेण नीरस्येवाभिभवः। तथा तमो अपि क्षीरवद्बलवदित्येवोच्यते। तर्हि दुर्बलस्य जगतः सर्गसमये अपि नोद्भवस्भव इत्यत आह तुच्छेन इति । आ समन्ताद्भवतीति *आभु *तुच्छ्येन । छान्दसो यकारोपजनः। तुच्छेन तुच्छकल्पनेन सदसद्विलक्षणेन भावरूपाज्ञानेन *अपिहितं छादितम् आसीत् । दधातेः कर्मणि निष्ठा । दधातेर्हि । 'गतिरनन्तरः' इति गतेः प्रकृतिस्वरत्वम्।

The universe lay powerless at the instance of its birth. It was surrounded by the insignificant. Ignorance was that insignificant principle.

*एकम् एकीभूतं कारणेन तमसाविभागतां प्राप्तमपि तत्कार्यजातं *तपसः स्रष्टव्यपर्यालोचनरूपस्य *महिना माहात्म्येन *अजायत उत्पन्नम् । तपसः स्रष्टव्यपर्यालोचनरूपत्वं चान्यत्राम्नायते 'यः सर्वज्ञः सर्वविद्यस्य ज्ञानमयं तपः' *(मु.उ १.१.९)* इति॥

Tapas is equal to knowledge.

kāmastadagrē āmavartatādhi manasō rētaḥ pratamaṃ yadāsīt

satōbandhumasati niravindan hṛidi pratīśyā kavayō maniṣā

कामः । तत् । अग्रे । सम् । अवर्तत । अधि । मनसः । रेतः । प्रथमम् । यत् । आसीत् ।

सतः । बन्धुम् । असति । निः । अविन्दन् । हृदि । प्रतिSइष्य । कवयः । मनीषा ।।

In the beginning, Love arose, which was the primal germ cell of the mind.

The seers, searching in their heart with wisdom, discovered the connection of the Being in Non-being. A crosswise line cut the Being from the Non-being.

ननूक्तरीत्या यदीश्वरस्य पर्यालोचनं जगतः पुनरुत्पत्तौ कारणं तदेव किंनिबन्धनमित्यत आह कामस्तदग्र इति। *अग्रे अस्य विकारजातस्य सृष्टेः प्रागवस्थायां परमेश्वरस्य मनसि *कामः *समवर्तत सम्यगजायत । सिसृक्षा जातेत्यर्थः।

The desire to create arose in the mind of the supreme being.

ईश्वरस्य सिसृक्षा वा किंहेतुकेत्यत आह मनस इति । *मनसः अन्तःकरणस्य संबन्धि वासनाशेषेण मायायां विलीने अन्तःकरणे समवेतत् । सामान्यापेक्षमेकवचनम्। सर्वप्राण्यन्तःकरणेषु समवेदमित्यर्थः। एतेनात्मनो

गुणाधारत्वं प्रत्याख्यातम्। तादृशं् *रेतः भाविनः प्रपञ्चस्य बीजभूतं *प्रथमम् अतीते कल्पे प्राणिभिः कृतं पुण्यात्मकं कर्म *यत् यतः कारणात् सृष्टिसमये आसीत् अभवत् ।

The seed for the desire was the impression of creatures from a previous cycle of existence.

भूष्णु वर्धिष्णवजायत परिपक्वं फलोन्मुखमासीदित्यर्थः। तस्यां च जातायां स्रष्टव्यं पर्यालोच्य ततः सर्वं जगत् सृजति। तथा चाम्नायते 'सो अकामयत बहुः स्यां प्रजायेयेति स तपो अतप्यत स तपस्तप्त्वेदं सर्वमसृजत यदिदं किंच' (तै.आ ८.६) इति श्रुतिः। आत्मनेत्थमवगभिते अर्थे विद्वदनुभवमप्यनुग्राहकत्वेन प्रमाणयति सत इति।

The supreme self-created many beings after deliberating.

*सतः सत्वेन इदानीमनुभूयमानस्य सर्वस्य जगतः *बन्धुं बंधकं हेतुभूतं कल्पान्तरे प्राण्यनुष्टितं कर्मसमूहं *कवयः क्रान्तदर्शना अतीतनागतवर्तमानाभिज्ञा योगिनः *हृदये निरुध्दया *मनीषा मनीषया बुध्दय। 'सुपां सुलुक्' इति तृतीयाया लुक् । *प्रतीत्य विचार्य । 'अन्येषामपि' इति सांहितिको दीर्घः । *असति सद्विलक्षणे अव्याकृते कारणे *निरविन्दन् निष्कृष्यालभन्त। विविच्याजानन्नित्यर्थः।

Beings experience the universe because of the element of sat within them. The root cause of sat is a remnant of karma at the end of the last kalpa cycle.

tirascīnō vitatō raśmirēśām adasvidāsīt uparisvidāsīt

rētōdā āsan mahimāna āsan svadhā avastāt prayati purastāt

तिरश्चीनः। विSततः। रश्मिः । एषाम् । अधः । स्वित् । आसी3त् । उपरि । स्वित् । आसी3त् ।

रेतःSधाः। आसन् । महिमानः । आसन् । स्वधा । अवस्तात् । प्रSयतिः । परस्तात् ।।

A crosswise line cut the Being from the Non-being. What was described above it, what below? Bearers of seed there were and mighty forces, thrust from below and forward move above.

एवमविद्याकामकर्मणि सृष्टेर्हेतुत्वेनोक्तानि। अधुना तेषां स्वकार्यजनेन शैघ्र्यं प्रतिपाद्यते। येयं नासदासीदित्यविद्या प्रतिपादिता यश्च कामस्तदग्रे इति कामो मनसो रेतः प्रथमं यदासीदिति यत्कर्म *एषाम् अविद्याकामकर्मणां वियदादिभूतजातानि सृजतं *रश्मिः रश्मिसदृशो यथा सूर्यरश्मिः उदयानन्तरं निमेषमात्रेण युगपत् सर्वं जगत् व्याप्नोति तथा शीघ्रं सर्वत्र व्याप्नुवन् यः कार्यवर्गः *विततः विस्तृतः *आसीत्। स्विदासीत् इति वक्ष्यमाणमत्रापि संबध्यते । विचार्यमाणानाम्' *(पा.सू ८.२.९७)* इति प्लुतः। तत्रोदात्तः इत्यनुवृत्तेः स चोदात्तः।

Actions related to creation spread out like the brightness of the Sun at sunrise.

*स्वित् इति वितर्के । स कार्यवर्गः प्रथमः किं *तिरश्चीनः तिर्यगवस्थितो मध्ये स्थित आसीत् किंवा *अधः अधस्तात् *आसीत्। आहोस्वित् *उपरि उपरिष्टात् किमासीत्। 'उपरि स्विदासीदिति च' *(पा.सू ८.२.१०२)* इत्यनुदात्तः प्लुतः।

What set of actions were at the beginning, the middle and the last?

आत्मन आकाशः संभूत आकाशाद्वायुर्वायोरग्निः *(तै.आ ८.१)* इत्यादिकया पञ्चमीश्रुत्या तत उद्गातारं ततो होतारमितिवत् क्रमप्रतिपत्तौ सत्यामपि

विद्युत्प्रकाशवत् सर्गस्य शीघ्रव्यापनेन तस्य क्रमस्य दुर्लक्षणत्वादेतेषु त्रिषु स्थानेषु प्राथम्यं कुत्रेति विचार्यते । एवं नाम शीघ्रं सर्वतो दिक्षु सर्गो निष्पन्न इत्यर्थः। एतदेव विभजते ।

No order was obvious because actions spread out rapidly like a lightning strike.

सृष्टेषु कर्येषु मध्ये केचिद्भावाः *रेतोधाः रेतसो बीजभूतस्य कर्मणो विधातारः कर्तारो भोक्तारश्च जीवाः *आसन् अन्ये भावाः *महिमानः। स्वार्थिक इमनिच् । महान्तो वियदादयो भोग्याः *आसन्। एवं मायासहितः परमेश्वरः सर्वं जगत् सृष्ट्वा स्वयं चानुप्रविश्य भोक्तृभोग्यादिरूपेण विभागं कृतवानित्यर्थः। अयमेवार्थस्तैत्तिरीयके 'तत् सृष्ट्वा तदेवनुप्राविशत्' *(तै.आ ८.६)* इत्यारभ्य प्रतिपाद्यते।

Beings (rētōdhā) who were the result of the seeds of karma, formed one part of this creation. The root elements (mahimāna) formed the other part.

तत्र च भोतृभोग्ययोर्मध्ये स्वधा । अन्ननामैतत् । भोग्यप्रपञ्चः *अवस्तात् अवरो निकृष्ट आसीत्। *प्रयतिः प्रयतिता भोक्ता *परस्तात् पर उत्कृष्ट आसीत् । भोग्यप्रपञ्चं भोक्तृप्रपञ्चस्य शेषभूतं कृतवानित्यर्थः। विभाषा परावराभ्याम्' *(पा.सू ५.३.२९)* इति प्रथमार्थे अस्तातिः। अस्ताति च' *(पा.सू ५.३. ४०)* इत्यवरशब्दस्यावादेशः। अवस्तादिति संहितायाम् ईषा अक्षादित्वात् प्रकृतिभावः॥

The remnant after the creation of the enjoyers became the objects of enjoyment. Svadhā connected the two.

kō addhā aēda kaḥ ihaḥ pravōcat kutaḥ ājātā kutaḥ iyaṃ visṛiṣtiḥ
ārvāgddēvā asya visarjanēnāthakō vēda yataḥ ābabhūva

कः । अद्धा । वेद । कः । इह । प्र । वोचत् । कुतः । आऽजाता । कुतः । इयम् । विऽसृष्टिः ।

अर्वाक् । देवाः । अस्य । विऽसर्जनेन । अथ । कः । वेद । यतः । आऽबभूव ।।

Who really knows? Who can presume to tell it?

When was it born? Whence issued this creation? Even the Gods came after its emergence.

Then who can tell from whence it came to be?

एवं भोक्तृभोग्यरूपेण सृष्टिः संग्रहेण प्रतिपादिता। एतावद्वा इदमन्नं चैवान्नादश्च सोम एवान्नमग्निरन्नाद' *(*श.ब्रा १.४.२१३*)* इतिवत् । अथेदानीं सा सृष्टिर्दुर्विज्ञानेति न विस्तरेणाभिहितेत्याह को अध्देति । * कः पुरुषः *अध्दा पारमार्थेन *वेद जानाति। *कः वा *इह अस्मल्लोके *प्र *वोचत् प्रब्रूयात् । *इयं दृष्यमाना *विसृष्टिः विविधा भूतभौतिकभोक्तृभोग्यादिरूपेण बहुप्रकारा सृष्टिः *कुतः कस्मादुपादानकारणात् । *कुतः कस्माच्च निमित्तकारणात् *आजाता समन्ताज्जाता प्रादुर्भूता । एतदुभयं सम्यक् को वेद को वा विस्तरेण वक्तुं शक्नुयादित्यर्थः।

What is the material source (upādhāna) of the creation and what is the causal source (nimitta) of the diverse creation? Clay is the material source of pot, but the potter is its causal source.

ननु देवाः अजानन्तः। सर्वज्ञास्ते ज्ञास्यन्ति वक्तुं च शक्नुवन्तीत्यत आह अर्वागिति।

All knowing Devas do not know the cause of diversity because they came later.

*देवाः च *अस्य जगतौ विसर्जनेन *वियदादिभूतोत्पत्त्यनन्तरं विविधं यद्भौतिकं सर्जनं सृष्टिस्तेन *अर्वाक् अर्वाचीनाः कृताः। भूतसृष्टेः पश्चाज्जाता इत्यर्थः। तथाविधास्ते कथं स्वोत्पत्तेः पूर्वकालीनां सृष्टिं जानीयुः। अजानन्तो वा कथं प्रब्रूयुः । उक्तं दुर्विज्ञानत्वं निगमयति। *अथ एवं सति देवा अपि न जानन्ति किल । तद्व्यतिरिक्तः *कः नाम मनुष्यादिः *वेद तज्जगत्कारणं जानाति *यतः कारणात् कृत्स्नं जगत् *आबभूव अजायत ॥

The creation from out of the root elements occurred before the birth of the Devas.

iyaṃ visṛṣtiḥ yataḥ ābabhuva yadi vā dadhē yadi vā na
asyādhyakṣaḥ paramē vyōman sa anga vēda yadi vā na

इयम् । विऽसृष्टिः । यतः । आऽबभूव । यदि । वा । दधे । यदि । वा । न ।

यः । अस्य ।अधिऽअक्षः । परमे ।विऽओमन् । सः ।अङ्ग । वेद । यदि । वा । न । वेद ।।

That out of which creation has arisen, whether it held it from or it did not; He who surveys it in the highest heaven, He surely knows or maybe He does not!

उक्तप्रकारेण यथेदं जगत्सर्जनं दुर्विज्ञानं एवं सृष्टं तज्जगत् दुर्धरमपीत्याह इयमिति। *यतः उपादानभूतात् परमात्मनः *इयं *विसृष्टिः विविधा गिरिनदीसमुद्रादिरूपेण विचित्रा सृष्टिःआबभूव आजाता सोऽपि किल *यदि *वा *दधे धारयति *यदि *व *न धारयति। एवं च को नाम अन्यो धर्तुं शक्नुयात् । यदि धारयेदीश्वर एव धारयेन्नान्य इत्यर्थः।

The causal source of the universe is the universal self. The universal Self is the only one capable of holding creation in place.

एतेन कार्यस्य धारयितृत्वप्रतिपादनेन ब्रह्मण उपादानकारणत्वमुक्तं भवति । तथा च पारमार्षं सूत्रं 'प्रकृतिश्च प्रतिज्ञादृष्टान्तानुपरोधात्' (वे.सू १.४.२३) इति । यद्वा । अनेनार्धर्चेन पूर्वोक्तं सृष्टेर्दुर्ज्ञानत्वमेव द्रडयति । को वेदेत्यनुवर्तते । इयं विविधा सृष्टिः यत आबभूव आ समन्तादजाययेति को वेद । न कोऽपि । नास्त्येव जगतो जन्म न कदाचिदनीदृशं जगदिति बहवो भ्रान्ता भवन्त्यपि। यतः। 'जनिकर्तुः प्रकृतिः' (पा.सू १.४.३०) इत्युपादानसंज्ञायां पञ्चम्यास्तिसल्। यस्तात् परमात्मन उपादानभूतादाबभूव तं परमात्मानं को वेद । न कोऽपि।

No one knows the supreme self who is the causal source of this diverse creation.

प्रकृतितः निमित्तभूतो वा जगज्जन्मेति हि बहवो भ्रान्ताः। तथा स एवोपादानभूतः परमात्मा स्वयमेव निमित्तभूतोऽपि सन् यदि वा दधे विदधे इदं जगत् ससर्ज यदि वा न ससर्ज । असंदिग्धे संदिग्धवचनमेतच्छास्त्राणि चेत्प्रमाणं स्युरिति यथा । स एव विदधे। तं को न वेद। अजानन्तो अपि बहवो जडात् प्रधानादकर्तृकमेवेदं जगत् स्वयमजायतेति विपरीतं प्रतिपन्ना विदधतो विधानमजानन्तोऽपि। स एव उपादानभूत इत्यपि को वेद । न कोऽपि। उपादानादन्यः तटस्थ एवेश्वरो विदधे इति हि बहवः प्रतिपन्नाः। देवा अपि यन्न जानन्ति तदर्वाचीनानामेषां तत्परिज्ञाने कैव कथेत्यर्थः।

Nature (Prakṛti) is the material source of diversity. Some are deluded with the idea that the universe arose on its own. They do not know the supreme self to be the causal source of the universe.

यद्येवं जगत्सृष्ट्रत्यन्तदुरवबोधा न तर्हि सा प्रमाणपध्दतिमध्यास्त इत्याशङ्क्य तत्सस्वभाव ईश्वमेव प्रमाणयति यो अस्येति । *अस्य भूतभौत्तिकात्मकस्य जगतः *यः *अध्यक्षः ईश्वरः *परमे उत्कृष्टे सत्यभूते *व्योमन्याकाशे आकाशवन्निर्मले स्वप्रकाशे । यद्वा ॥ अवतेस्तर्पणार्थात्

अन्येभ्यो अपि दृश्यन्ते' इति मनिन् । 'नेड्वशि कृति' इतीत्प्रतिषेधः। 'ज्वरत्वर' इत्यादिना वकारोपधेयोः ऊट् । सप्तम्या लुक् । 'न ङिसंबुध्दयोः' इति नलोपप्रतिषेधः॥ व्योमनि विशेषेण तृप्ते। निरतिशयानन्दस्वरूपे इत्यर्थः। यद्वा। अवतिर्गत्यर्थः। व्योमनि विशेषेण गते व्याप्ते । देशकालवस्तुभिरपरिच्छिन्न इत्यर्थः। अथवा । अवतिर्ज्ञानर्थः। व्योमनि विशेषेण ज्ञातरि विशेष्टज्ञानात्मनि। ईदृशे स्वात्मनि प्रतिष्ठितः। श्रूयते हि सनत्कुमारनारदयोः संवादे 'स भगवः कस्मिन् प्रतिष्ठित इति स्वे महिम्नि' *(छा.उ ७.२४.१)* इति ।

The supreme Self resides in his own space which is filled with peace and knowledge. It is not tainted by objects and spatial divisions and time divisions.

ईदृशो यः परमेश्वरः *सो *अङ्गः। अङ्गेति प्रसिध्दौ । सो अपि नाम *वेद जानाति। *यदि *वा * न *वेद न जानाति। को नाम अन्यो जानीयात्। सर्वज्ञ ईश्वर एव तां सृष्टिं जानीयात् नान्य इत्यर्थः॥

The supreme Self is the only one who knows this creation. He is called popularly by the name anga.

9. Conclusion

There is a close similarity between the primordial plasma of the physicist and the primordial energy per the cognition of Rishis. An understanding of the close relationship between primordial energy and Cosmic Consciousness is essential to learn about the three kinds of Space in Vedic literature. It explains names such as Brahman, Hiranyagarbha, Prajapati and Purusha in mantras.

Primordial energy or Svadhā is the material cause of this sentient and inert creation. It exists in the space of Cosmic Consciousness. An atom is a wave of electromagnetic energy per quantum theory. Energy alone powers thoughts in the brain. The Space of Cosmic consciousness permeates into physical space and the space of thoughts which are perishable.

We learn of the origin of the cosmic mind and of individual beings from the Nāsadīya sūktam. We learn of the origin of Devas and of the material universe from the Ambhasya pārē sūktam. Mantras only mention the distinction between the cosmic mind and cosmic consciousness. Yoga Vaśiṣṭā explains the distinction in detail. Vaśiṣṭā's parables bring us the understanding that consciousness permeates Void which per theoretical physicist is filled only with Vacuum energy.

Rishis used different names to distinguish configurations in the space of consciousness. These distinctions can not be understood without knowing the way in which primordial energy differentiated during creation. Energy differentiations create subatomic particles and atoms from plasma per modern science. Yoga Vaśiṣṭā clarifies the role of energy differentiations from macro-cosmic and microcosmic viewpoints. It explains how two other kinds of space will collapse if differentiations disappear from the field of energy of cosmic consciousness.

Anyone attempting to understand the logic behind (a) the phrase "Devas are a part of a cosmic being" (b) the distinction between Devas and their Shaktis (c) the liberation of an ātmā and (d) the worship of agni must explore ideas in the Nāsadīya sūktam and the Ambhasya pārē sūktam further. Yoga Vaśiṣṭā is an excellent source for anyone wanting to explore the idea that consciousness is the cause behind creation.

Native experts published a comprehensive commentary on all the Vedas in the fourteenth century under the guidance of Sāyaṇācārya. Sāyaṇācārya's treatment of the Nāsadīya sūktam and the Ambhasya pārē sūktam is unique. He explains meditative insights from eons ago against contemporary view of cosmology. He shows how a phrase in a mantra may point to an entirely different section in the Vedas. A good interpretation of mantras must rely on a knowledge of such hidden references and on alternate ways to find the meaning of words in mantras.

One of the mysteries of human consciousness is its power of cognition. Meditators tapped into this power eons ago to gain insights into the workings of the subtle universe. They recorded them as mantras. Technology has now matured enough to bring new insights into the workings of the physical universe. Definite similarities exist between these two insights. This book presents the extent of similarities between ancient and modern Cosmologies and brings out the complementary nature of cognition and technology.

www.ingramcontent.com/pod-product-compliance
Lightning Source LLC
Chambersburg PA
CBHW051808170526
45167CB00005B/1924
* 9 7 8 1 5 4 4 2 2 3 9 8 8 *